Y0-BCV-821

The
Systems Analyst

The Systems Analyst

How To Design Computer-Based Systems

JERRY W. ATWOOD

Zounds Consulting Company

HAYDEN BOOK COMPANY, INC.
Rochelle Park, New Jersey

Library of Congress Cataloging in Publication Data

Atwood, Jerry W
  The systems analyst.

  Includes index.
  1. Electronic data processing. 2. Electronic digital
computers. 3. System analysis. I. Title.
QA76.A844   003   77-404
ISBN   0-8104-5102-6
ISBN   0-8104-5101-8   pbk.

4	5	6	7	8	9	PRINTING
79	80	81	82	83	84	85	YEAR

Introduction

In the past few years, there has been some debate as to what the term Systems Analysis actually means. Several books have come across my desk all with the title of Systems Analysis: one a very advanced treatment of higher mathematics; another a detailed explanation of simulation and operations research; and several on computers. All of these books seemed well-written and sincere in their treatment of Systems Analysis, but none of them seemed to be written about the actual job of the person called the systems analyst. This book is about that person.

Systems analysis may be applied to almost any area and has been used in aerospace engineering, military defense, and many other industries; however, the job position of systems analyst is a computer industry term. As this book describes, this is the person who analyzes data and information from the point where it is created to its final objective use. Almost always the systems analyst is concerned with the use of a computer to produce the final result.

This book was written to describe the job of the systems analyst and the methods, principles, and techniques the analyst uses to accomplish his tasks.

Although the book should be of some use to computer programmers who desire to move up to the position of systems analyst, it is primarily designed as a textbook for an introductory course in systems analysis, as taught in the department of management, data processing, and computer science.

Prerequisite to this course is usually either Introduction to Data Processing/Computer Science or Computer Programming, preferably both.

Because of the prerequisites, this book is concerned mostly with the general principles of analysis and design of systems that will use a computer and not with the functions of hardware, programming, and such standard subjects as binary codes, all of which should be covered in a good introductory course before this book is used.

The study of this book leads into the higher courses of information systems, business systems, data base management, and advanced subjects of systems analysis including selection of equipment and telecommunications systems.

Several problems are included which allow the student to attempt the actual work of the systems analyst.

The text is designed so that the chapters should be studied in sequence, as each chapter leads into the next. It is sometimes a good idea to assign students a term problem at the beginning of the class to be completed as the chapters are presented. Therefore, the systems flowcharts could be drawn when Chap. 3 is completed and output designed after Chap. 6, etc. When this method is used, sometimes one class per week is assigned as a work period to allow students to obtain some experience in working in teams.

The following is a suggested lecture outline for a 30 classroom meeting term. Longer terms should be filled in with team projects, problems, and case studies.

CLASSROOM MEETING

1 SYSTEMS—Introduction to systems and their parts. Defining the system and the work of the systems analyst.

2 SYSTEMS—Explanation of the work of computer systems and method of using them. Detailed explanation of the stages of development of a computerized system.

3 THE SYSTEMS STUDY—Explanation of the origination of systems studies, the appointment of teams and project leaders, and use of the master plan. Define the user and other workers who come in contact with the systems team.

4 THE SYSTEMS STUDY—The steps to a systems study. Examination of management needs and need for information in an organization. Understanding of the term "data" and how it is created.

5 SYSTEMS FLOWCHARTING—Presentation of flowcharting symbols and the use of the flowchart template. Difference between systems flowcharting and program flowcharting.

6 SYSTEMS FLOWCHARTING—Work in the preparation of systems flowcharts using the symbols for magnetic tape, disk, computer program, and manual methods.

7 SYSTEMS FLOWCHARTING—Flowcharting a system from the beginning where data is created to the end where information is produced.

8 SYSTEMS DESIGN—Introduction to the phases of systems design and the use of computers. Study of the types of design.

9 SYSTEMS DESIGN—The steps to design of a system. Study of the design of functions, limitations of design, and the need for standards during design work. Considerations of personnel and manual steps in the system.

10 SYSTEMS DESIGN—Completing the design of a system. Possibilities of automating some functions. The design freeze and presentation to management. Review of steps to systems design.

11 DATA MANAGEMENT—Explanation of a data management system and the considerations of the system during design phases. The organization of data into files and records.

12 DATA MANAGEMENT—Methods of processing data, standardization, security, and control. Types of files encountered in data management.

13 DATA MANAGEMENT—The approach to file design, selection of storage media, and processing methods. Definition of data sets and data codes.

14 OUTPUT—Design of various types of output and business reports. Specifying output on layout sheets. Some classroom work in using file and output layout sheets.

15 OUTPUT—Techniques of controlling output and insuring accuracy of output. Explanation of the data reduction section.

16 OUTPUT—Use of forms, design of preprinted output forms, and realtime output. Use of feedback, and the difference in systems output and application output.

17 INPUT—Considerations to designing input, the steps of input design, and the creation and capture of data.

18 INPUT—Survey of various methods of preparation of data. Specifying data items and use of input layout sheets. Record and file specification.

19 INPUT—Survey of the input editing techniques, controlling accuracy of input and the documentation of input items.

20 PROCEDURES AND LOGIC—Introduction to the programming cycle and the relationship of the systems analyst to the computer programmer. Development of logic for systems.

21 PROCEDURES AND LOGIC—Error checking by programmers, types of processing, and logic for output reports. Manual tasks in a system.

22 CONTROL—Steps to general control and what control should accomplish in a system. The concept of feedback and the monitoring of a system.

23 CONTROL—Introduction to control as applied to input, processing, output, and hardware. Special controls by systems codes and computer operator.

24 DOCUMENTATION—Explanation of the importance of documenting the system and having standards for documentation. Explanation of the systems specifications.

25 DOCUMENTATION—Introduction to the five basic documents that should be present in each system, the systems specifications, the equipment specifications, the programming documentation, the operation run book, and the users manual.

26 ECONOMIC CONSIDERATIONS—Survey of the costs in a computer system and cost control.

27 ECONOMIC CONSIDERATIONS—The terms for acquiring computer equipment and some of the economic principles involved.

28 USE OF TELECOMMUNICATIONS—Definition of a transmission system and how telecommunications can be used in a system. Considerations for systems design using telecommunications.

29 USE OF TELECOMMUNICATIONS—Understanding transmissions, channels, services, and the definition of terminals.

30 IMPLEMENTING A SYSTEM—Ordering equipment, selection of personnel, training needed, etc. Testing the system, file conversion and Systems Implementation.

In summary, the future looks bright for the profession of the systems analyst. With computers taking more and more of a daily role in the life of our society, there will always be a need for the person who can analyze the requirements for using a computer.

JERRY W. ATWOOD

Memphis, Tennessee

Contents

The
Systems Analyst

1

Systems

The advancement of mankind throughout history has been achieved by ideas, inventions, and methods which usually increased the complexity of life. In the last century this has probably been more true than at any other time. Many marvelous devices have been developed which extend physical and mental abilities while government and business processes produce more goods and services than ever before. All of the benefits we enjoy today are made possible only by the functioning of thousands of components that interact to carry out the tasks required in a modern society. Men, machines, and resources are required to be in the proper place at the proper time, each contributing to the progress and maintenance of our way of life.

As a task becomes more complex, decisions about the accomplishment of that task become more difficult to make. This problem was recognized as most important during World War II when the managers of the military effort faced complex problems of transportation, supply, personnel, communications, production, and, of course, obtaining money to pay for solutions to these problems. These managers began to seek better ways to analyze their problems, and from their efforts a new approach to finding solutions was slowly developed.

Even before World War II scientists were beginning to discover that all objects and functions have certain common characteristics. These characteristics are the same for simple biological cells as they are for complex administrative problems. Several researchers such as Ludwig Von Bertalanfly and H. Aiken began to predict, and rightfully so, that a study of these characteristics would have wide implications in all areas of knowledge. For example, if the human body has the same characteristics as a machine, perhaps something could be learned about the human body that could be applied to the machine and vice versa.

This almost universal set of characteristics began to be called a system. The analysis of an object, a task, a function, or a situation by examining its basic characteristics is usually called the systems approach.

The word system is a much used term. In today's complex society, there are accounting systems, biological systems, computer systems, management information systems, military systems, engineering systems, and ecological systems. The list is practically endless. Throughout man's history there have always been systems. The Egyptians needed engineering systems to construct the great pyramids. The Romans

1

had very complex military systems. Many countries had transportation systems, especially shipping. It is only recently, however, that the word system has come into general use. It seems that almost anything that functions or accomplishes an objective can be thought of as a system, and this is essentially what the systems approach is all about. It is a way of thinking.

DEFINITION OF A SYSTEM

A system has been defined in many ways, such as:

1. An orderly arrangement
2. A logical order
3. A mathematical method used to study objects
4. A logical method of doing something

Probably the most widely accepted definition of a system, however, is simply that it is a group of parts which form a whole. The definition of a system which is used in this book is the following:

A system is a group of interrelated parts, elements, processes, components, functions, etc., which together accomplish some specific objective.

Generally, the word analysis is defined as the task of examining the parts which make up a whole. Therefore, systems analysis is the task of examining the parts, elements, processes, components, functions, etc., which make up a system so as to determine their relationship to each other and how each contributes to the accomplishment of the objective of the system. It follows that system design is the act of improving a system, or in the case of a new system, the specification of the parts which will make up the system.

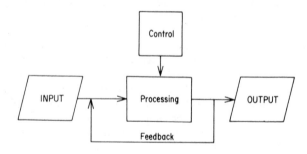

Fig. 1-1   Basic model of a system and its five parts

A system, shown in diagram form in Fig. 1-1, consists of five basic parts, as follows:

1. Input

2. Processing
3. Output
4. Control
5. Feedback

These parts of a system can be defined as follows:

1. *Input:* Anything which enters the system.
2. *Processing:* Any action upon the input.
3. *Output:* Anything that leaves the system, that is, whatever is produced by the processing of the input.
4. *Control:* The direction or adjustments of the processing.
5. *Feedback:* A measurement or indication of the quality of output.

Analyzing natural and man-made systems by the study of input, processing, output, control, and feedback has been applied to widely diversified areas of science. Biologists, electrical engineers, mathematicians, economists, and many others have begun to apply the systems approach to their areas of knowledge. However, this book is not about systems in general but rather only about systems which use digital computers to produce information. Therefore, the five parts of a system are redefined as follows so as to be applicable to "information systems":

1. *Input:* Digits or characters, usually called data, which enter the system in raw form. Input will consist of the numeric digits 0 through 9, the alphabetic characters A through Z, and any special characters allowed in the system, such as these: . , + * $, etc.
2. *Processing:* Any action upon the data, such as addition, subtraction, multiplication, division, expansion, deletion, etc.
3. *Output:* Information, but more specifically that information that meets the needs of the users of the system.
4. *Control:* The direction or adjustments of the processing which can be a computer program, written specifications of how tasks are to be performed, management policies, government regulations, etc.
5. *Feedback:* Any measurement or indication of the quality of the information being produced by the system.

The use of the systems approach, or systems concepts, has evolved because of the complexity of business, government, and society. It has been estimated that 29 percent of the gross national product of the United States is spent on information processing. This includes machines, supplies, and manpower to process raw data into information.

As the magnitude of the task of processing information expanded, new technology was developed which would allow business, government, and other areas of society to accomplish its objectives. Probably the most important of these developments was the digital computer. However, as the digital computer began to be used, managers soon discovered that there was much more to the enterprise than simply installing a machine. A careful study of requirements, objectives, and methods was needed before a computer system could begin to perform as expected. This study of the objectives of computer use and the design of the methods of use has been called systems analysis and design. The aim of this book is to present the techniques that have been developed for using computers and the factors taken into consideration before any system is implemented.

IMPACT OF COMPUTER SYSTEMS ON ORGANIZATIONS

All organizations that make use of computer systems discover benefits and problems which they have never faced before. Below is a listing of just a few of these benefits and problems:

1. *Use of data never before available:* A computer system, through its speed and ability to process data, makes information available that has never before been available to management. For example, daily inventory status and other such information is almost immediately at hand, thereby optimizing the use of material and resources.

2. *Centralized files:* Introduction of computer systems allows centralization of all data created and used by the organization. Access to this data is becoming easier each year as technology progresses. Instant information is becoming a new concept.

3. *Quality of data or information:* Many organizations find that the quality of information they receive improves with the use of the computer. Through techniques presented in later chapters, it can be seen that accuracy of information can be controlled within an information system.

4. *Ability to monitor:* The managers of an organization find that computer systems allow them to monitor the actual performance of their organization far better than they have ever been able to before. Timely, accurate, and quality information contributes to better decisions.

5. *Wider span of control:* Organizations find that the computer helps them in controlling more functions than ever before. The monitoring techniques mentioned above allow one manager to control more personnel and more functions than previously possible.

6. *Reduced time needed to carry out functions:* Many clerical functions can be carried out by the computer system, reducing the time and manpower needed to accomplish a task. This frees many workers from boring, routine jobs and allows more time for analysis of the organization.

7. *Attention to exceptions:* The computer system has the ability to find exceptional conditions within the system and report them. When such exceptions are reported, they can be corrected before they seriously affect the organization.

Although these are just a few of the benefits of using a computer, probably none of them will be realized unless the system is well planned. The analysis and design of a system which will be efficient, meet objectives, and provide benefits is the job of the systems analyst.

DUTIES OF THE SYSTEMS ANALYST

Generally, the systems analyst is expected to analyze and design systems of optimum performance. That is, they must produce at the lowest cost an output which fully meets management objectives. The analyst usually works on one project at a time.

He does not have routine duties but is rather concerned with different areas and different applications at various times according to which project he is assigned. The job description of a systems analyst would be quite varied; however, some duties can be noted, as follows:

1. He is a consultant to management on computer systems planning, organizing, operation, and control.
2. He is an analyst of problem areas and makes recommendations for improvement.
3. He evaluates present systems and looks for areas where improvements can be made.
4. He designs new systems or redesigns old systems and prepares the systems specifications in detail.
5. He analyzes requirements for equipment and makes recommendations.
6. He prepares documentation, develops standards, and produces special management reports in the information processing area.
7. He must keep himself well informed on advanced techniques, such as operations research and statistics, which could be applied to improve systems.
8. He must have a good understanding of the duties of those who work with him, such as computer programmers, computer operators, and project managers.

Systems analysis is the investigation of a system with the objective of improving it. The investigation is the analysis of the system, and the design is the improvement of the system. The analyst is also often called upon to create new systems. This is one reason why systems analysis can be thought of as both an art and a science.

CHARACTERISTICS OF THE SYSTEMS ANALYST

It would be very difficult to develop any formula for determining what persons would perform well as systems analysts. General observations of analysts have shown that a good systems analyst will have the following characteristics:

1. He has a creative mind.
2. He has the ability to think logically, symbolically, and abstractly.
3. He works well as part of a team.
4. He has a good knowledge of the capabilities of computer equipment and computer software.
5. He works within the constraints of time tables and schedules and is conscientious in meeting deadlines.
6. He seeks the ideas of others and involves them in the analysis and design of a system.
7. He considers himself both a teacher and a student, willing to train others but also constantly keeping himself up-to-date through night classes, personal reading, and short seminars.
8. He listens attentively but does not depend upon the opinions of others, rather determining for himself what the real facts are.
9. He is familiar with organizations and especially the principles of business and government administration.

A good analyst is very conscious of the fact that systems are created by and used by human beings. Installing a machine does not make a system. The user of the system should have the ultimate satisfaction, and the final criterion is that a system meets the objectives of the user. Therefore, the systems analyst shows consideration for the people affected by the system design. He is systems-oriented not machine-oriented.

The best way to introduce a highly automated system to an organization so that there is a minimum of psychological problems and personnel adjustments is to involve everyone possible in the analysis and design of the system. This means that a constant interchange of ideas and suggestions should take place between the systems analyst and the users, managers, and workers who are affected by the system.

WHAT COMPUTER SYSTEMS DO

Computer systems have the ability to accomplish various functions, as follows:

1. They collect and convert data into machine-readable form.
2. They store data in a computer file medium.
3. They process data by computer instructions.
4. They retrieve data from a computer file medium.
5. They display data in the required format or medium.
6. They transmit data to other locations.

Not all of these functions are accomplished by either one program or one machine. The system is the method of accomplishing these tasks, from the beginning, where the data is collected until the end, where data becomes the objective output.

Collection and conversion of data: Data must be collected and converted into machine-readable form. This is usually accomplished by writing or printing a source document or by some action of a data collection device. A source document is any written data on any medium, usually paper, from which it can be collected and made ready for conversion into machine-readable form. Common examples of source documents are invoices, time cards, and texts. These documents are converted by some means into a computer format. There are many ways to make this conversion, the most common being the key punch with which an operator punches into cards the data which the document contains. Other methods of performing this task are optical scanning, mark sensing, MICR, etc. All of these methods will be explained later.

Storage of data in computer file medium: Once data is converted into a computer-readable form, it must be stored on some medium which the computer can access. Punched cards can be accessed through the card reader. However, some types of systems require large storage capacity in some computer file medium. Examples of these file media are magnetic tape and magnetic disk.

Processing of data by computer instruction: Information or data is processed by a computer by means of various instructions which the computer can carry out. Calculations such as addition, subtraction, multiplication, and division change the data into different forms. Editing, transferring, expansion, and deletion are all tasks which the computer can perform.

Retrieval of data from the computer file media: Once data is processed or while it is being processed, it must be retrieved to be presented for output. Types of file media from which data can be retrieved are magnetic disk and magnetic tape.

Display of data in required format or medium: The objective of an information system is to produce some type of required output. The most common output is the printed page—reports, invoices, payroll checks, etc. However, there are many other types of output which are mentioned later in this text, such as microfilm, punched cards, paper tape, display screens, etc.

Transmittal of data to other locations: Once data is in some machine form it can be transmitted to various locations remote from the actual location of the data. Most of this transmission is made by the use of telephone lines and is called telecommunications. Such systems require special consideration on the part of the systems analyst who designs them.

METHODS OF USING COMPUTER SYSTEMS

Many different methods of using a computer have evolved throughout its short history. These methods of using a computer can be listed as:

1. Program run
2. Batch processing
3. Integrated systems
4. Real-time systems
5. Total systems
6. Network systems

All computers work under the direction of a set of instructions called the program. In modern computers, programs are usually called steps, meaning that each program performs one task or step toward the accomplishment of some objective. The accomplishment of this objective is usually called a job. Therefore, several programs or steps make up one job. The relative complexity of a job, a step, a program, or a system essentially determines the difference in the method of using the computer system.

Program run: Early model computers worked under the control of a single program which essentially could be thought of as a one-time run; that is, the program run used one set of input and produced one set of output. Most of these programs were run one at a time and usually accomplished only one specific objective, such as computing a statistical average or producing a simple report. As one program was started and other programs finished, there was a stopping period before the run and a stopping time after the run. The inefficiency of running a single program at a time was soon realized by the managers and users of computers.

Batch processing: As time progressed, it was found that single program runs could be made more efficient (use less processing time) by batching them together. It was also found that input data could be batched into groups, thereby making its control much easier and the processing more efficient. This method of using the computer became known as batch processing. In other words, both the data and the programs using the data were grouped into batches. The processing was reasonably automatic with no starting or stopping times between the different programs; however, there was starting and stopping time between the batches.

Integrated systems: The batch processing method developed into what is known today as integrated processing systems. Integrated systems are essentially the continuous running of programs, steps, and jobs and the continuous processing of input and output. This continuous processing method has been called the job stream. The jobs beings processed could be considered as subsystems of a greater system which would achieve the objectives of information processing. In many cases, the output of one computer program became the input of another program. Each program or step within this system was interrelated with other programs or jobs that were being run by the computer. As the tasks of a system became more and more interrelated—for example, the output of a payroll system might become the input into an accounting system and the accounting system might be related to production scheduling, inventory control, and sales reports—it became more and more apparent that analysis, planning, and design of programs, tasks, procedures, and methods were needed. Managers became aware that programs and systems could not be developed to stand alone, but that the development of each program and system must consider other programs and systems within the organization which might be interrelated with them. To add to these problems of analysis and design it was also discovered about this time that computers could process more than one program at the same time. These techniques of sharing the processing capability of the computer became known as time sharing, multiprogramming, and real-time systems.

Real-time systems: A computer system which is capable of producing output immediately in response to an inquiry is called a real-time system. The key to understanding which systems could be called "real-time" is the factor of immediate response. If information is received fast enough to affect a decision or the environment, it could be called immediate response; however, in modern computer systems this speed should not exceed two or three minutes. In many cases, it should be less than a second.

Examples of real-time systems are reservation systems, control systems, and some management systems which use a computer. Response to a request for a hotel reservation that was received in a minute or two could be called immediate response, whereas data to control the flight of a space ship heading for the moon might need to be much less than a second to be called an immediate response.

A real-time system will consist of a central computer and files, a communication link with that computer, and a terminal of some type to send and receive information. This configuration could depend upon manual operation at the terminal or it could be completely automated with the terminal sending and receiving signals based upon some type of sensing device.

Real-time systems are possible because of the advanced methods of using a computer, such as the integrated processing capability, the technical ability to run more than one program at a time, and the use of telecommunications for sending and receiving data from remote locations. The analysis and design of real-time systems requires some special attention to programming methods and the equipment to be used. Chapter 12 presents more detailed information.

Total system: As it became possible to use computers in more advanced ways, it was apparent that complete total systems could be developed. A total system is also sometimes called a management information system. In theory, everything within the organization would be connected to a computer. Every time a transaction took place in the organization it would enter the computer system at the point of creation. Any information needed from the system would be immediately available. Many organizations today are working toward total-management information systems, although few actually have been achieved.

Network systems: Real-time systems and some integrated systems use telephone lines for sending and receiving data. As it was found that data could be sent economically in many types of applications, users of computer systems began to theorize about the possibility of network systems which would provide not only information within an organization but also outside it. The network system is a method of using a computer which is developing now and which may be prevalent in the future. In such a system not only do total systems exist for each organization involved, but they are also tied into large, complex networks of computers connected together by telephone lines. Information becomes available to anyone using the system, much like a public utility provides telephone service today. What will result from the development of large networks in the years ahead is not predictable at this time; however, there is little doubt that the systems of the future will be multicomplex, more interrelated, and made up of many subsystems probably connected together by telecommunications capabilities. For example, the income tax return of a large corporation can now be filed on magnetic tape with the Internal Revenue Service. Some day, a corporation's computer will simply dial up the Internal Revenue Service and directly send the data needed for an income tax return to the IRS computer. As these systems develop, their planning and analysis become more important not only because of the interrelations within a company, but also because of the company's relationships with other organizations.

System analysis and design techniques developed over the last few years allow users of computers to approach the design of their systems scientifically. It has been stated that the '60s were the age of the computer programmer; most probably the '70s and '80s will be the age of the systems analyst.

DEVELOPMENT OF A COMPUTERIZED SYSTEM

The steps usually followed in developing a digital computer system can be listed as follows:

1. Systems analysis
2. Feasibility study
3. Systems design
4. Equipment selection
5. Programming
6. Conversion
7. Documentation
8. Implementation of the system

Not all these steps take place in the order given. In fact, many are performed at the same time. Phases in the development of a digital computer system can be defined briefly as follows:

1. *Systems analysis:* The task of investigating the system as it is presently operating and determining the desires, needs, and objectives of the persons using the system.
2. *Feasibility study:* The task of studying the economics, technical aspects, and operational requirements of a system. The feasibility study essentially tries to answer the three questions: Is this system technically feasible? Is it economically feasible? Can it be feasibly operated?

3. *Systems design:* The specification of what the characteristics of a system will be. In other words, determining the exact requirements, specifying the tasks to be performed, determining the best way to perform the tasks, and establishing required performance levels.

4. *Equipment selection:* Determining what equipment will best perform the tasks required of a system in terms of efficiency and economy.

5. *Programming:* The writing of the instructions which the computer will carry out.

6. *Conversion:* Converting an old system to a new system or beginning to use a completely new system.

7. *Documentation:* The recording of the specifications of a system such as the objectives, the equipment specifications, the program listings, and the manuals on how to use the system.

8. *Implementation of the system:* The management, maintenance, and operation of the system after it has begun to be used.

Entire books have been devoted to each of the above subjects, and more research, standardization, and proven techniques are needed in all areas. The study of systems analysis sometimes is frustrating because in many areas such as the selection of equipment, there are no definite answers to problems, only considerations. As with the application of all types of knowledge, systems analysis and design techniques must be adapted to the problems and objectives of organizations according to past experience and to the particular application which is to be implemented.

SUMMARY

There is little doubt that the society in which we live is becoming more complex each day. Scientists began to discover that all objects or functions, simple or complex, have certain characteristics. This set of characteristics came to be called a system.

A system has been defined in many ways; however, probably the most widely accepted definition is simply that it is a group of parts which form a whole. The analysis of a system is defined as the task of examining the parts which make up the whole.

All systems consist of the following five basic parts: (1) input, that which enters the system; (2) output, that which the system produces; (3) processing, the actions upon the input; (4) control, the regulation of the processing; and (5) feedback, measurements upon which control is based.

Inputs to one system can be the output of another system, just as the output from a system can be the input into other systems. Therefore, the term integrated systems has come into use. There are several types of computer systems, most of which are information producing systems. An analyst designing a system must consider the inter-relation of all its parts and its relationship to other systems.

The development of computerized systems involves the following steps: systems analysis, feasibility study, system design, equipment selection, programming, conversion, documentation, and implementation. The systems analyst must work closely with the users and managers of a system for their approval of his design concepts.

The duties of a systems analyst are varied but basically can be said to be analysis and design, analysis being the act of investigating the system and design being the act of improving the system. A good systems analyst should be creative and have the ability to think logically, symbolically, and abstractly. It is probable that systems analysis is both an art and a science, and, therefore, that the systems analyst should use all techniques of science that are available to him but must supply the art by his own creativity.

Questions

1. Name and discuss the five parts of a system.
2. Given the definitions and discussion in this chapter, would you consider the human being a system? Why?
3. Discuss the manufacturing of automobiles as a system.
4. Is systems analysis and design an art or a science? Give reasons to support either answer.
5. Name some specific benefits which might be realized by the installation of a well-planned and designed computer system in a large department store chain.
6. Name three duties of the person whose job title is "systems analyst."
7. What characteristics and educational background would you look for if you were hiring a systems analyst?
8. It can be stated that the systems analyst is interested in the accomplishment of all tasks from the beginning to the end. In an information system, what is the beginning and what is the end?
9. What method of using a computer is most prevalent today? What will it be in the future?
10. Discuss in brief, general terms the systems development of a university class registration system using a computer.

Exercises

1. Given the following example of a university considered as a system:

   Input: Untrained students.
   Processing: Teaching of students.
   Output: Trained graduates.
   Control: University regulations, administration, and faculty.
   Feedback: Student's grades, performance on job, complaints, etc.

   make a list of five systems that you are familiar with. In general terms, state what these five systems consist of by analyzing their five basic parts: input, processing, output, control, and feedback.
2. Consider the following situation: A mining company obtains iron ore from the earth and ships it to a steel mill. The steel mill processes the iron ore and markets the finished metal to a typewriter manufacturing plant. The typewriter manufacturing plant produces portable typewriters for the consumer market. Using a block diagram such as the one shown in this chapter, draw these three systems and show how they are interrelated. At the same time label input, output, processing, control, and feedback. Notice how outputs of one system become inputs into another system.
3. You are a systems analyst hired by a toy manufacturer to advise them on the

possible use of a computer for accounting, inventory control, and production scheduling. The company has never used a computer before. The president of the company's first question to you is, "How will you approach the development of the system and what benefits or problems can be expected from it? Write a report to the president of the company explaining the phases of systems development which will be undertaken to implement the system. Finish your report by explaining some of the possible impacts of the computer in the toy manufacturer's organization.

2
The Systems Study

The systems analysis phase of systems development is called the systems study, the feasibility study, or simply the function of systems analysis. Before the design of a new system can take place, there must be an analysis of all the various factors which must be considered. The design of a new system involves the design of hundreds of elements—both qualitative and quantitative. There is no easy formula or checklist approach to making a systems study or for carrying out any phase of systems development. Each case will be different. However, there are several principles and concepts which should be considered in all types of systems analysis. They are presented in this chapter.

HOW SYSTEMS STUDIES ORIGINATE

A systems study originates with the recognition of a problem within an organization. Figure 2-1 diagrams the way in which a systems study might originate and how it might be carried out. This recognition may come from the lowest operating level or from the highest management level. Once the problem is recognized, there are usually many meetings, discussions, and committee recommendations as to whether it is a valid problem which needs solving, how the problem should be stated, what the objectives are, and when the problem should be solved. This interaction usually takes place between top or middle management personnel who would be affected by any new system and the systems personnel. This discussion of the problem, the decision to make the systems study, and the assignment of systems analysts should not be a haphazard operation. There should be written specifications available to all levels within the organization on how to originate possible systems studies, what approval is needed, and how decisions will be made.

Once it is decided that a systems study is needed, the analysis of the problem becomes a project. Normally, a project manager is assigned and team members then chosen. At this point, there is no commitment by anyone in the organization to either use the computer or to use any system which may be developed. It is simply a study of the problem area.

WHEN SHOULD A COMPUTER BE USED?

Some general characteristics calling for the use of computer systems processing are as follows:

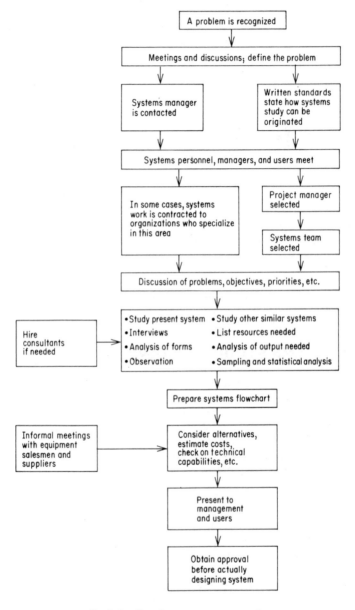

Fig. 2-1   Carrying out a systems study

1. When large volumes of information are involved
2. In complex situations
3. When cycles are involved, such as daily, weekly, or monthly cycles
4. In repetitive logic situations
5. When accurate solutions are possible

Large volumes of information: In any organization where large volumes of information must be processed, a computer is usually the first thought of most analysts since a computer has the ability to process large volumes of information rapidly and efficiently. Banks, insurance companies, and other organizations which generate much paperwork have found that the computer solves many of their clerical processing problems.

Complex situations: For problems which are solvable but so complex that they cannot be done without many calculations or much expenditure of manpower, the computer has been a great benefit. Examples are feed-mix problems, where a mill must make decisions on what feeds to mix into its product based upon current prices of feed, protein content, etc.

Cycles: Any processing which must be repeated daily or in some other recurring cycle can be automated with some benefit. One of the considerations of the analyst in all types of design should be to spread costs as widely as possible. When large volumes of information are to be processed in cycles which are repeated over and over, the cost of the system and the cost of developing and designing the system is greatly lowered. In other words, a system used daily has a small unit cost, whereas a system run only once a year has a high unit cost. Therefore, the principle is to spread the cost as widely as possible.

Repetitive logic situations: Computers can best handle tasks where the logic of processing is the same for each item. If each item entering a system must be processed differently, computer programming may become so complex as to be inefficient or uneconomical.

Accurate solutions: Use of a computer denotes that accurate solutions, or at least approximate solutions, are possible. Examples may be found in accounting, airplane reservations, and credit billing. In these systems it is possible to obtain valid output or answers. Situations where accurate solutions are usually not possible are predicting horse races, prices on the stock market, or commodity futures. Although a computer can be used to analyze such situations and come up with an optimum answer, normally it is used only in situations where solutions are practical.

MASTER PLANNING

Before an organization attempts to use a computer, a master plan for systems development should be decided upon by the top managers. Systems development in a piecemeal fashion usually cause problems. Some of the preplanning steps that should be taken before any systems studies take place are the following:

1. Define all the objectives of the organization.
2. List all the functional areas of the organization.
3. Make a list of all applications that have potentials for the use of a computer or automated equipment.
4. Decide upon the priority of applications.
5. Set up methods for review and appraisal of systems.

All objectives of the organization should be defined in both general and detailed specifications. The objectives help to decide what systems have priority. Many of them lend themselves immediately to systems study. After they have been decided upon, all functional areas within the organization should be listed to as low a level as possible. Shipping and receiving, purchasing, sales, market research, accounting, finance, and cost control all might be listed as functions of an organization. By exploring this list, it is possible to discover the areas which may need systems studies, as well as to understand the interactions between the areas.

Once the objectives and functions have been listed, it should be possible for management to determine which applications have potential for the use of a computer.

All of the above steps should be documented in writing. Some type of priority, moreover, must be placed upon the development of applications. Time tables should be prepared, implementation plans developed, a rough estimate of cost and benefits made, and expected impact upon the organization analyzed.

Finally, within these written specifications, there should be a method for review and appraisal of systems. In this master plan, the approvals needed to produce a systems study, use the computer, prepare budgets for making studies, and other considerations should all be stated in writing.

THE USER

The systems analyst is not the user of the system. He may develop the system and never come in contact with it again. Therefore, particular attention should be given to the users, that is, the persons who utilize the system's output. All steps in system development should be approved by the users as well as top management (Fig. 2-2). There must be allowance for interaction between the systems group and the users. Systems developed without the approval of the persons who will be using them are usually disastrous.

GENERAL STEPS OF A SYSTEMS STUDY

There is no easy formula or definite step-by-step approach to conducting a systems study or the analysis phase of systems development. Generally speaking, the systems analyst should consider the five parts of a system—input, processing, output, control, and feedback—as the basis for his study. In other words, some of the questions which the analyst will be asking about the system are: What kind of inputs enter the system? What output does the system produce? What controls can be applied to the system to assure quality of output and that the system meets its objectives? What kind of information can be used for feedback to assure that the system will operate as expected or can be adjusted if it does not operate as expected?

Although there are some scientific approaches to answering these questions, the systems analysis phase is probably a creative task as much as it is a scientific task. The systems analyst must not only look for answers to the above questions, he must also consider how the system will fit in with other systems within the organization. He must ask himself what subsystems are contained within the system which he is studying and whether or not an old system is in operation at the time of the systems study. In most cases, the functions of the new system are usually already being carried out in some other manner within the organization, probably by manual methods. In some cases the systems study will be for an entirely new system, and the creativeness of the systems analyst will be called upon even more.

Although there are no definite steps to conducting a systems study, some general tasks should be carried out before the study is considered to be complete. These can be listed as follows:

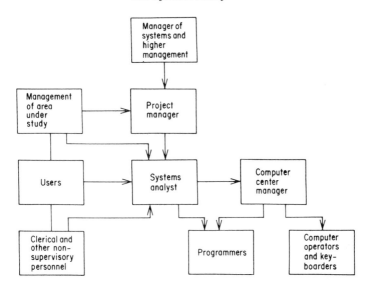

Fig. 2-2   Relationship of the systems analyst to various personnel during a systems study

1. Problem definition
2. Stating objectives of the system
3. Organizing for the systems study
4. Information gathering
5. Study of other systems
6. Interviewing
7. Analyzing documents
8. Inventory lists
9. Output analysis
10. Sampling and statistics
11. Observation of personnel
12. Preparation of systems flowcharts
13. Preparation of alternative ways for the system to function
14. Presentation of the systems study to management and users for approval
15. Obtaining approval from management before proceeding further

The first twelve steps above could be called the first phase of the systems study, upon which the later recommendations and preparation of alternatives are based.

Problem definition: It is most important that there be a written definition of the problem which initiated the systems study. This definition should be precise and agreed upon by all parties concerned, that is, by the systems team, the users of the system, and the managers of the system. Accompanying this definition should be written documentation of the assignment of the members of the systems team, as well as the selection of a project manager. Investigation of the problem and the carrying out of the system analysis phase will be based upon these specifications. General problem statements such as "Inventory costs are too high" should be made more specific by stating requirements, such as reducing the cost of carrying inventory, eliminating shortages in inventory, etc.

Stating objectives of the system: The objectives stated in the systems study should concern those areas with the greatest impact upon the organization in question. In a business situation, this usually means the area with greatest impact upon profit. The objectives of the systems study should be specific and in writing. Accompanying this statement should be a statement of the limits of the study and the conditions under which the study will be made, such as time constraint, cost budgets, etc. Limitations are sometimes called the boundaries of the system. Areas to be studied may also be stated so that the analysis is not too general and broad.

The objective of a system is its output; therefore, the output to be produced by the system should be stated as specifically as possible. It might be producing a daily inventory status report, a report on inventories which are too large or too small, a calculation of optimum inventory amounts, or an automatic output of purchase orders for those stock items which need to be reordered. Just as with the problem definition, the statement of objectives should be as detailed as possible and should be given a great deal of thought by the managers and users of the system.

Organizing for the systems study: After a clear problem definition has been developed and the objectives of the system are agreed upon, it is time to select the study team. In some organizations, study teams exist as a department of the organization; however, systems study personnel are usually selected from the various areas of an organization according to need. Sometimes consultants or specialists are hired on a short-term basis to help with the systems study.

The size of the systems team will probably always be greater than one. If only one person conducts the systems study, there cannot be that interaction of ideas which is needed to create a well-rounded system. In a large systems study the team may consist of as many as 10 to 20 persons, if not more. There is no definite rule as to what type of personnel should be selected; however, generally speaking, personnel should be about 40 percent functional, 40 percent systems oriented, and 20 percent specialists. That is, if a systems study is to consist of ten team members, then four of them should be specialists in the area to be studied, such as accounting, inventory, production, etc. Four others should be specialists in systems analysis and design. The last two could be specialists in equipment selection, economics, mathematics, statistics, or some other area which may be needed to conduct the study.

The team members may be full-time, part-time, or act in a consulting capacity. The persons who select the team usually base their selections on experience and education; however, the main objective is to have a well-rounded group. A project manager should be selected to act as coordinator between the users and managers of the system and the systems study team. It is usually a good idea to announce widely within the organization the selection of the team, their objectives, and their duties so that all personnel will be aware of the study being conducted.

Information gathering: The first step in a systems study should be to become familiar with the existing system, if there is one. Areas that should be investigated are the history and background of the organization, government regulations which affect the system, projections of the future volume, capacity, etc., of the system, and, in general, all information possible that bears on the system. This operation is essentially a fact-finding task. Many sources are used, including company files; however, the most fruitful are studying other systems, interviewing, analyzing documents, making inventory lists, analyzing output, sampling of statistics, and observing personnel. All these steps could be listed under the general heading, Information Gathering; however, they are discussed separately in the following paragraphs.

Study of other systems: No matter what system is being studied, it is likely that a similar system is in use at some other organization. In the case of a competitor, it is not always possible to visit the other organization; however, similar systems should be visited and analyzed by the systems team whenever possible. Also, all articles, books, etc., about systems similar to the one being studied should be obtained.

Interviewing: Probably the most common method of information gathering is the interview. Much useful information comes from interviewing the persons who will be managing the system and who will be using the system. Early in the systems analysis phase a list of persons who will be interviewed should be made. Before an interview takes place there should be a discussion of what information is needed, and the entire systems team should participate. Pre-written questions should be developed which will be asked the interviewee.

Interviewing different types of personnel means that different techniques of interviewing are necessary. However, in all cases the interviewer should always follow the rule of courtesy, which means that an appointment should be made for each interview, questions should be prepared in advance, and technical details, terminology, and "jargon" of the computer industry should be avoided. The interview should be conducted in an informal manner, and the systems analyst should realize that his job is mostly to listen, rather than to tell how things will be.

Written results of the interview should be produced and distributed to all members of the systems team. Personnel can be interviewed more than once by different members of the team or can be re-interviewed if need be by the same person.

When interviewing top management personnel, the systems analyst will be mostly concerned with broad policies and objectives. These interviews should be among the first to take place. Top management will have the final decision as to whether the system will be implemented and how it will operate. Knowing the thoughts of these managers can save many wasted hours interviewing other personnel.

Middle management interviews should be concerned with the functions of that particular area. Even if the system being developed is strictly an accounting system, there should be interviews with managers of sales, production, engineering, research, to see how the system might affect them and interact with their areas.

In interviewing nonsupervisory personnel, detailed analysis should take place. The intricacies of the system should be discussed at this time with the persons who will actually be using the system from day to day.

Analyzing documents: Every document used in the present system should be analyzed as to its content, usefulness, and need. Most documents will be redesigned for the new system, especially those documents which contain data needed by the computer system. The design of documents will be discussed later in this book.

Files should be kept of all documents used by the old system, and rough diagrams should be prepared of those documents to be used by the new system. These documents can then be referred to by members of the systems team at any time. Every person who uses a document should be asked about its possible redesign and its exact usefulness to the system. Of course, every unnecessary form should be eliminated and new forms should be created with discretion.

Inventory lists: In the initial phase of the systems study the systems analyst should prepare lists of all the resources of the system, for example, a list of all machines, a list of all personnel, a list of all procedures, and so forth. As the analysis progresses, lists of items not currently used but possibly needed by the new system should be prepared. Although equipment and software is not selected at this time, the systems team may wish to keep proposed lists of items of equipment, types of computer software, and types of file organization which may be needed in the future. When the actual selection phase takes place, these lists can be helpful. As with other documents used by the system, the inventory lists should be kept in a file folder and should be readily available to all persons on the systems team.

Output analysis: All outputs of the system should be analyzed, such as reports, special forms, and other types of output which perform some particular task objective. Since the output is the essential objective of the system, the first analysis to be undertaken should probably be concerned with what is to be produced. All managers, users, and members of the systems team should be asked to specify in detail the type of output they desire from the system. Users of the system are particularly concerned because the output of the system is basically what they use in their daily work. Although the output is not designed until the design phase, there should be detailed analysis of what is desired during the analysis phase. Careful analysis of the output desired can save much time and trouble in the analysis and design of other parts of the system, since all parts of the system are concerned with the basic objective of producing output.

Sampling and statistics: It is not always possible to investigate every item in a file or every factor which may affect a system. For example, in an inventory control system, it would probably not be possible to analyze every item in inventory. However, the analyst should consider the possibility that every tenth item or every thirtieth item, etc., might be analyzed, and from this sample of data, conclusions could be drawn about the make-up of the data, such as the average cost of items in inventory, average turnovers, etc.

Observation of personnel: In many cases the best way to gather facts about a system is to observe the personnel while they are at work. By noting the procedures needed to perform a function in a manual system, the analyst can gain much information which can be used for the cost estimates and time estimates that must be applied to a computerized operation. Job descriptions, operating manuals, and other such texts can also help the systems team understand the manual functions of a system. If job descriptions have to be rewritten for the new system, the personnel manager or the person involved should be consulted.

Preparation of system flowcharts: Once the analyst has begun to understand how the old manual system works, he may prepare flowcharts of this system and then revised flowcharts to show how a new system might operate. These flowcharts or diagrams map the input, processing, and output of data as it flows through the organization. (A simple systems flowchart is shown in Fig. 3-9. Standard symbols used in flowcharting and the methods of flowcharting are explained in detail in the next chapter.)

Preparation of alternative ways for the system to function: Once the systems team agrees that the analysis of the system is complete, it must determine how the new system should operate. Various alternatives should be investigated, such as comparing a manual system with a fully computerized system, or a manual computer system with a fully computerized system.

The preparation of such alternatives is probably one of the most difficult tasks the analyst will face and is also the task where the most creativeness is needed. Of course, technical knowledge of the capabilities of computer equipment and computer programming is important. However, general concepts of how the system should operate are also important. A detailed design of the system is not necessary at this time. Therefore, statements of alternatives should be general in nature so as to be easily discussed by the systems team, the managers, and the users. They should present concepts of how the system will operate rather than exact details of its functioning. A typical description of alternatives might read as follows:

The proposed inventory control system will produce basically one inventory report daily. This report will show current levels of stock of all items in inventory. An alternative report will indicate items which need special attention, such as those which have fallen below desired levels of stock or which are over-stocked. To produce the daily inventory report, it is envisioned that receiving slips from the warehouse will be prepared for the computer daily and matched against purchase orders and usage of stock, which are reported as they occur. This type of system will allow projecting current levels of stock, stock that is on order, and stock that has been used. It is envisioned that the system will use two magnetic disks for a master inventory file and the data input transactions will be prepared by punch card. From the systems flow chart it can be seen that two computer programs will be needed to produce this report, one of them to update the master file and the other one to actually print-out the report.

Such a description as this allows for general flexibility in the discussion of a system. If possible, estimates of computer time needed, preparation costs, and other costs which might be incurred by the system can be presented; however, a detailed cost analysis is not necessary at this time. As many alternatives as are practical should be offered to management and users. In fact, one of the alternatives would probably be that the manual system should be retained. The advantages and disadvantages of each alternative should be listed.

Presentation of the systems study to management and users for approval: The presentation of a study should be as professional as possible. The systems team should rely on salesmanship to present their ideas in the most complimentary manner so that they will be digested, analyzed, and acted upon favorably by management. Therefore, the presentation should be well thought out in advance. Use of visual aids and any other methods which might improve the presentation are recommended. Written copies of the study should be presented to each member of the management team and those users most affected by the system.

Obtaining approval from management before proceeding further: The final task of the systems team is to obtain approval of their work before proceeding into any detailed design of the system or detailed study of the feasibility of the system. Many times the feasibility study will be incorporated into the systems analysis phase and presented at the same time as the systems study. It is, of course, not desirable to design the system in detail until it is determined that it is feasible. Feasibility of a system should be determined in three ways: (1) technically, (2) economically, and (3) operationally.

The question of technical feasibility is not very difficult to answer. What can be done technically can usually be determined by referring to manufacturers' specifications. Most computer salesmen will be able to answer the question of technical feasibility with ease. Therefore, the problem is not technological in most cases but rather economic, that is, will a system be worth the cost?

The most difficult question to answer, however, is whether the system will work as expected, that is, will it be feasible to operate? In most cases, the answer depends upon the quality of the system's design.

In summary, the objective of the feasibility study is to establish justifiable reasons for implementing a system.

Management's decision to implement a system is based upon (1) the advantages and disadvantages of each alternative, (2) the return on investment or cost savings from implementing the system, and (3) nonmonetary advantages which might incur from installing the system. Once management has selected one of the alternatives, time tables should be developed to ascertain whether the system can be implemented at the assigned time within the stated budget.

Managers, of course, may base their decision upon as little or as much information as they think they need. Some managers will feel confident in their decision with minimal systems analysis and only rough cost projections. Other managers will desire details about all costs which might be incurred and assurance that the system will work as expected. Of course, the type and size of system being installed and the extent to which computers are already used by the organization will help determine the kind of feasibility study required for management to feel confident in its decision.

MANAGEMENT CONSIDERATIONS

Many writers have affirmed that a successful system is the direct result of involving management in the systems analysis phase. Since management has the final decision, it, of course, cannot be ignored. Care indeed must be taken to discover exactly what it desires. The desires of management may be listed as the following:

1. Control
2. Speed and efficiency
3. Volume processing
4. Accuracy
5. Cost benefits

Control: Control over all functions of an organization is a prime management objective. Conditions that are not functioning as they should must be brought to management's attention and adjusted. This subject will be discussed in a subsequent chapter.

Speed and efficiency: Management is very concerned with deadlines, time constraints, etc. The system must produce what is desired within acceptable time limits. In many cases, efficiency depends upon time limits and vice versa.

Volume processing: Volume processing can best be achieved if the outputs are spread out through the cycle of processing; that is, the system should not produce all output at the end of the month. Increased production at the same cost is always a welcome benefit.

Accuracy: Accuracy is very important to management. Exception reporting is also to be considered by the analyst. Figures that are erroneous or out of limits should be flagged on reports so that they will not be overlooked. Later chapters discuss this subject in more detail.

Cost benefits: The basic consideration in the mind of every manager is that of cost. Essentially, computer systems have high fixed costs and low variable costs. Therefore, volume is the key to lower costs. The analyst must be very concerned about costs and should constantly ask himself if there is a less expensive way of doing things. Of course, there are many trade-offs of costs: quality vs. inexpensiveness, etc. The analyst is concerned with producing adequate records for cost control, developing cost standards, use of budgets and control of budgeting, cost accounting and reporting, and periodic auditing. A cost benefit is a monetary saving or increase in income created by some decision or systems design characteristic such as reduction of personnel, increased sales, elimination of duplication, etc. There are only two ways in which a cost benefit can be obtained—by decreasing cost or increasing income. The systems analyst should try to make his system do both.

MANAGEMENT FUNCTIONS

As the analyst progresses with the systems study, he must pay attention to the functions of management. These functions are planning, organizing, motivating, communicating, controlling, and decision making.

In the planning phase, management is concerned both with predicting and controlling the performance of the organization in the future. A systems design should contribute to both short-range and long-range planning. Information generated by the system should be usable for predicting future courses of action.

A system should provide for good organization within all functional areas and promote their efficiency. Although systems functions may cross over organizational lines, they should do this easily and without increasing costs.

Managers must be concerned with motivating personnel to produce their most efficient work, and the design of a system must consider such motivation. Systems procedures should not be overly tedious, psychologically harmful to anyone, or so constraining that they allow little room for creativeness.

One of the basic needs of an organization is to communicate. Communications is the heart of many systems. They must be so designed that each person will be suitably informed, perhaps automatically, by the system.

A principal management function is controlling. Systems design must help the manager control the organization through reports and timely information.

The final management function is decision making. This subject will be discussed in later chapters.

REPORTING NEEDS OF AN ORGANIZATION

To make good decisions a manager must have information upon which to base them. Types of reports needed by an organization can be classified as: (1) timely information reports, (2) required reports, (3) transaction handling, and (4) reference reports.

Timely information reports: These are reports upon which decisions are based. They can be detailed in nature, summaries, or involve exception reporting. Whatever their nature, the analyst must make sure that they adequately meet the needs of the manager and the users of the system. To ensure this goal, interviewing about their format and constant interactions between the user and the systems group are needed.

Required reports: Every organization must produce a certain amount of required reports. These are reports which must be submitted to the government, or they may be characteristic of the type of organization, such as invoicing.

Transaction handling: Reports are needed to keep track of the many transactions which take place within an organization. For example, a list of every customer who has placed an order could be produced as a subsystem of a system which handles orders and sales. In many cases, such reports are integral parts of the system.

Reference reports: Many reports are simply detailed registers of status and characteristics of an organization. These are reports such as listings of all items in inventory, lists of all customers who have done business with the organization, etc.

In all the reporting needs of an organization, the analyst should strive for simplicity in design and be very careful to eliminate unnecessary reports and keep overlapping to a minimum.

DATA CREATION

Computer systems depend upon the use of data created either within or outside the system. Data is created whenever any transaction takes place or when objectives are carried out. For example, data is created when a physical inventory is taken, when a sale is made to a customer, when new products are produced on an assembly line. All of this data might be of use to the system. It is the decision of the systems analyst and the systems designer as to what data is needed to produce valid and useful output.

Data can be defined simply as any character or group of characters such as an alphabetic character, a number, or a special symbol which has some meaning. Data can also be said to be a picture of a certain state or status of an organization. In the chapter on data management, data is discussed in detail, including its structure as used by a computer system.

SUMMARY

The conducting of a systems study can be a long and tedious operation. It is probably the portion of the systems development which requires the most creativeness because the information gathered must be turned into a working system. The systems group has the difficult job of looking for long-term benefits from a system that must also take care of short-range problems. It is important that the study team concentrate on the total information system. It should plan for any exceptions which might arise in the system and give due consideration to time constraints and requirements.

A system may be designed from scratch or it may merely be a redesign of an old system. Before any systems study takes place, a master plan should be drawn up to establish its methods as well as priorities for applications of the computer.

To conduct a systems study, a clear problem definition is needed. Objectives and limitations of the system must be stated. Selection of the systems study group must be based upon these requirements. As the systems analyst studies the system, he must consider management needs, constraints, and functions. The functions may be described as planning, organization, motivation, communication, control, and decision making.

As the study progresses, the team will become more and more familiar with the present system. It will have made an analysis of output reports and will have come to understand all aspects of the system, including the volume and frequency of data, peak loads, etc. It will have prepared inventories of items of present equipment used and personnel needed within the system. It will have identified present problem areas and determined how present procedures might be improved. It will have made a rough estimate of the performance of the present system and analyzed its organizational structure. It will have prepared work-flow diagrams of the present system, which are simply diagrams of how work is accomplished in the organization.

After the system has been studied in detail, its feasibility must be determined and management approval obtained before proceeding any further. At the same time, care must be taken to obtain the approval of the users of the system.

Questions

1. List some characteristics of a bank which would lead you to believe that it could make use of a computer.
2. A book arrives at your desk entitled, *A Master Plan for Systems Development.* What would you expect to find in such a book?
3. Who is the user of a system? How does the systems analyst interact with the user?
4. Discuss four ways of studying a system.
5. Name three techniques of interviewing.
6. Discuss how you would interview the following persons: (a) the president of a company, (b) the accounting manager, (c) a purchasing agent, and (d) a typist who prepares billing statements.
7. Discuss what a manager does and what his concerns are.
8. The State Industrial Commission is interested in installing a computer to collect statistics on industry and business. Reports are to be produced for a wide variety of projects, including pollution control, usage of resources, tax and revenue planning, and public relations. The State Commissioner has selected you as project manager and authorized you to select seven systems analysts from among state employees. What type of individuals would you select for your systems team and why?

3
Systems Flowcharting

Systems flowcharting is a method of describing a system in diagram form so as to simplify and more clearly depict the operations involved. Systems flowcharts are used for two basic reasons: (1) to communicate ideas of systems processing to other individuals and (2) to confirm the analyst in his beliefs about the processing methods to be used.

Flowcharts are used for communication between two or more individuals. Referring to a flowchart is the clearest method of discussing a system. Points of interest can be localized since input and output are depicted in pictorial form. Moreover, although a system may be clear in the mind of the analyst, a flowchart is a sort of proof that his ideas will work. By diagraming his system, the analyst can see where inputs enter it, where outputs leave it, and what is necessary to produce the inputs and outputs. He can see clearly what programs and other procedures will be needed to make the system function.

Sometimes during, or at the completion of, the systems analysis phase, the analyst should prepare a formal flowchart. Although he may have used rough diagrams of the system previously, a formal flowchart is needed before a system design can begin. It is the basic document to be presented to management and users that describes how the system will work.

DEVELOPMENT OF FLOWCHARTS

Several mathematicians, among them John Von Neuman, proposed early in the age of computers the use of flowcharts to describe the logic of processing. Although the symbols they developed for this purpose were different from those used today and their methods differed widely, they founded the basis for flowcharting. Since then, the American National Standards Institute (ANSI) has established certain symbols which are used within the industry. In the last few years, the ANSI symbols have been made to conform to the specifications of the International Standards Organization (ISO). Consequently, flowchart symbols are standard throughout the computer industry all over the world, with a few exceptions.

The objective of this chapter is to present methods for describing a system by means of flowcharts and to show how flowcharts reflect the logical processing which takes place within a system. We will be concerned with only one of the two types of flowcharts—the systems flowchart. The other type of flowchart is known as the logic flowchart or program flowchart and will be discussed in a subsequent chapter. The systems flowchart is not used for depicting the detailed logic of processing but rather for showing where data will be stored, the type of storage medium which will be used, where the input enters the system, where the processing steps take place (not specifically what the processing steps are), and where the output is produced.

FLOWCHART PREPARATION

A systems flowchart is usually prepared with the help of a flowchart template. Templates come in various sizes and shapes. Although the symbols that appear on them must conform to basic geometric shapes, there is no standardization of their size. The symbols are divided into three categories: (1) basic symbols, (2) symbols related to programming, and (3) symbols related to systems. All symbols are presented in Fig. 3-1, but we will be concerned here only with the systems symbols. Note that some symbols are composed of two or more other symbols.

When a flowchart is prepared, the flow of processing is presented to conform to the dictates of reading a printed page, that is, from left to right and from top to bottom. In most cases, a flowchart will begin at the top left-hand margin and end somewhere near the bottom, right-hand corner of the page, but this arrangement will vary according to the needs of the system.

BASIC FLOWCHARTING SYMBOLS

Process: The rectangle symbol denotes a processing function. It represents a group of logical operations that will take place to accomplish a single major function. For example, a flowchart symbol for producing inventory status would simply be a rectangle with perhaps the words "Inventory Status Run" written inside of it.

Annotation: Any comment or annotation which might make a flowchart clearer can be placed on it by using an open rectangle connected to the flowchart wherever desired by a dotted line. The comment or annotation is written within the rectangle and along the dotted line.

Input-output: Input-output procedures are depicted by a parallelogram. Normally, in a systems flowchart this input-output symbol is replaced by a symbol that shows the medium which the input or output file is placed upon, such as punched cards, magnetic tapes, magnetic disks, etc. Although the parallelogram can be used in systems flowcharting, it is normally used in a logic flowchart or programming flowchart.

Connector: The connector symbol, a circle, is used whenever it is necessary to exit from a page or from one part to another part of a flowchart. If the names of files are used over again when depicting the input-output of files, there is no need to use a connector. Although a basic symbol, it is used mostly in program flowcharts. A rectangle with a triangle below it is a special off-page connector symbol which is used in some specialized cases.

Flowlines and arrows: Symbols in a flowchart are connected by flowlines and arrowheads. The arrowhead denotes the direction of the flowchart, that is, the flow of data from input to output. In some special cases arrowheads are not used. However, it is a good habit to use arrowheads on all flowcharts, even when the logic or processing depicted is very simple and would not be confused. Without arrowheads it is sometimes difficult to tell which file medium is input and which is output. The

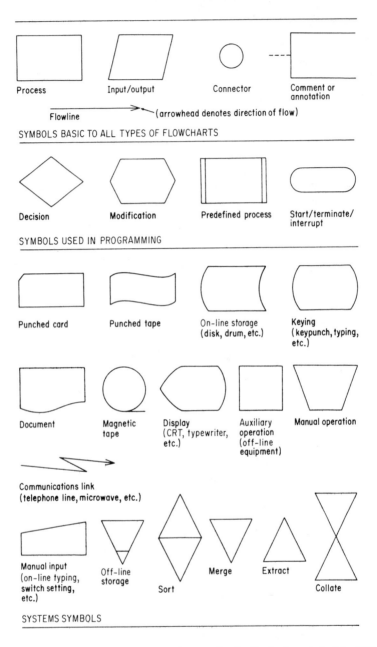

Process   Input/output   Connector   Comment or
                                                           annotation

Flowline   ───►──(arrowhead denotes direction of flow)

SYMBOLS BASIC TO ALL TYPES OF FLOWCHARTS

Decision   Modification   Predefined process   Start/terminate/
                                                                     interrupt

SYMBOLS USED IN PROGRAMMING

Punched card   Punched tape   On-line storage   Keying
                                            (disk, drum, etc.)   (keypunch, typing,
                                                                     etc.)

Document   Magnetic   Display   Auxiliary   Manual operation
                  tape   (CRT, typewriter, operation
                                   etc.)   (off-line
                                                     equipment)

Communications link
(telephone line, microwave, etc.)

Manual input   Off-line   Merge   Extract
(on-line typing,   storage
switch setting,   Sort   Collate
etc.)

SYSTEMS SYMBOLS

Fig. 3-1   Flowchart symbols (all standard except for the "keying" symbol of the IBM
           Corp.)

arrowhead is normally placed at the end of the flowline next to the symbol; however,
arrowheads can also be placed in the middle of the flowline. If flowlines cross, it does
not mean that they are connected. However, crossing flowlines should be avoided if

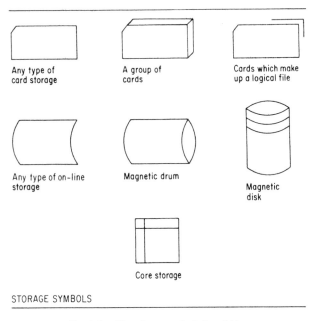

Fig. 3-1   Flowchart symbols (cont'd.)

possible since they are often confusing and make for an unorganized flowchart. If two flowlines meet, this means that the incoming flowline joins the flow at this particular point. Arrows should be used where flowlines join. In a systems flowchart every symbol must be joined by a flowline.

SYSTEMS FLOWCHARTING SYMBOLS

When a device is connected to the central processing unit of a computer, it is considered to be on-line. When it is not connected to the central processing unit (or when a procedure takes place outside the computer system), it is considered to be off-line. There are systems flowchart symbols to depict both on-line and off-line operations. Generally, the symbols denote types of data carriers or data mediums, that is, mediums which can hold or transmit data to the central processing unit in computer readable form. Thus the analyst must have some rough idea of the file medium upon which the data will be placed. If the data file mediums are unknown or not understood, however, the general input-output symbol can be used. Later, during the systems design phase, the file medium can be decided upon and a systems flowchart can be drawn to depict the different files.

Punched cards: The punched card symbol is a rectangle with the corner notch resembling the physical appearance of an actual card. A single symbol denotes all types of cards, regular, standard, 80-column punched area, etc. For normal use, moreover, this symbol can be used to denote an entire deck of cards. However, a card deck symbol is also available which denotes a collection of punched cards. Also, a specialized symbol which repeats the upper right corner of the punched card symbol can also be used to denote a group of punched cards.

In a systems flowchart it is usually a good idea to denote the procedures for creating the punched card rather than start the system with the punched card as unexplained input.

Documents: The document symbol denotes any written information on paper. This can be a source document from which data is collected, or it can be the final output document printed by the computer system. The document symbol should be used to specify all paper used in the system, such as invoices, bills of lading, inventory sheets, output reports, listings, accounting registers, payroll checks, special printed forms, etc.

Magnetic tapes: The circle with a small line below it to the right is the symbol for magnetic tape files which are used for input or output to a system. Magnetic tape is a medium upon which data can be magnetically recorded and stored.

Punched tapes: Data stored in punched paper tape is denoted by a curved rectangle. In today's computer industry, paper tape is usually used for data communications when data is collected over long-distance telephone lines.

On-line storage: Any input-output storage connected to the central processing unit can use the on-line storage symbol. However, this symbol most often denotes the use of a magnetic disk or drum. Since the drum is no longer in common use, the symbol will most likely denote a disk.

Display: The display symbol indicates an output of the system which is not retained in written form but rather shown temporarily by cathode ray tube or other type devices. In some cases the display symbol can also be used to denote specialized types of output such as a digital plotter page or perhaps the output from a console typewriter.

Manual input: The manual input symbol is used to denote input by any manual means such as keying into a typewriter on-line to a computer, setting sense switches or dials, etc.

Manual operation: The manual operation symbol is used for depicting off-line processes, those which take place not at machine speed but usually at human speed. Such operations can be a physical review of data cards, obtaining a sheet of paper from a file, or any other process which is not connected to a machine.

Auxiliary operations: A square is the symbol for auxiliary operations which are similar to manual operations except that they use equipment not connected to the computer system. The addition of a batch control total by a ten-key adding machine or separating carbon paper from output printed forms would be examples of this type of operation.

Merge: The combining of two or more files into one is called a merge. The merge operation symbol is the triangle.

Extract: The extract is the opposite of the merge, the separation of one file into two or more files. The extract symbol is the same as the merge except it is upended.

Collate: Two merge symbols meeting together form a collate symbol. Collate is very similar to merge except two files are brought together under some specific logic rule, such as one card from one file placed between six cards from another file.

Sort: The sort is a basic operation in most computer systems. Sorting is based on a key, that is, an identification number such as a social security number, invoice number, customer number, credit card number, or other such identifying number. These keys identify records and for most types of processing are sorted in ascending sequence, that is, from the lowest number to the highest number.

Communications link: The jagged flowline which is similar to a depiction of a lightning bolt is the symbol used to depict transfer of data by some type of communication link. These communications link symbols can flow in either direction and usually enter programs from some type of terminal or possibly from another computer program or computer operation. The subject of communications links will be developed more fully later.

PREPARATION OF FLOWCHARTS

Finding a starting point is a basic problem with a flowchart. Some persons in the field advocate starting with the final output document and working back to the beginning source document. Others recommend the reverse. Some analysts believe that the preparer of the flowchart should start with the part of the system that is clearest and most familiar to him. From this he can determine how the inputs are prepared and what output is produced and then expand the flowchart. In the last analysis, the way the flowchart is prepared depends upon what is needed by the system, the type of logical processing that will take place, and the method of collecting and presenting the information.

The simplest example of a systems flowchart is one in which one input produces one output. For example, Fig. 3-2 denotes card input and printer output. This could be a simple listing of data contained upon cards or could be the processing of some type of report from cards. Another simple example is a card-to-tape type of processing. Figure 3-3 depicts an operation where cards enter a system and are placed upon magnetic tape, perhaps in an edited or card-image form. It is possible to expand these charts backwards to the data collection source or forward to present still more detailed information. For example, Fig. 3-4 depicts the operation of preparing the cards from the original document, producing the cards, sorting the cards into some type of sequential order, and putting the final punched cards onto magnetic tape.

Systems flowcharts essentially depict a logic of "input, process, and output." The output from the process either becomes the input into another process, or it serves as the final output. In Figs. 3-2, 3-3, and 3-4, each processing symbol comes between the input symbol and the output symbol. Building these processing modules, interrelating them, and tying them together will present the flow and procedures of the system.

Fig. 3-2   Flowchart depicting printing of data from cards

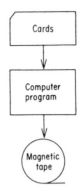

Fig. 3-3   Flowchart depicting data from cards being transferred to magnetic tape

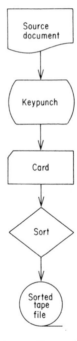

Fig. 3-4   Flowchart for preparing, sorting, and storing data on magnetic tape

OPERATIONS OF A SYSTEM

Information systems consist of basic operations which can be depicted by a systems flowchart. The most basic operation is the update. Files are not static. Information is constantly changing and must be updated. The file to be updated is called the master file, and the information or data which updates the file is called the transaction file. Figure 3-5 shows a simple update program. Notice that a new file is

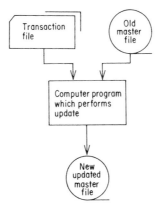

Fig. 3-5   Flowchart for updating a file

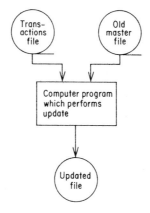

Fig. 3-6   Flowchart for updating a magnetic tape file

created from two files—the master file and transaction file. Figure 3-6 shows the same operation, but one using magnetic tape alone. The merge and sort operations are shown in Fig. 3-7. The files enter the program and are denoted as sorted or merged output.

Operations may involve random (direct access) processing (Fig. 3-8A) or sequential processing (Fig. 3-8B). If the operation involves both an input and an output to the direct access device, then arrowheads on the flowlines are shown in both directions, as shown in Fig. 3-8A.

Programs often produce outputs which become inputs to another program. Figure 3-9 shows program runs producing two outputs—one output becoming the input to the next program, the other output being another desired result. Data communications lines are used in Fig. 3-9 to show collection of data from remote points.

A basic type of run is editing for control of errors. In Fig. 3-10, the data enters the system, is edited, and if it is not correct, is returned to the beginning of the system. The feedback is denoted by a broken line. In processing cycles where a tape is updated

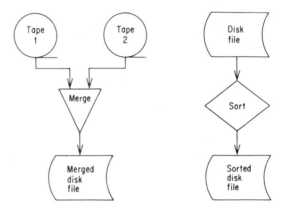

Fig. 3-7   Flowchart for merging two magnetic tape files and sorting the disk file

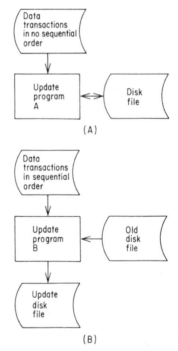

Fig. 3-8   (A) Flowchart for random processing to update file in same disk space, and (B) flowchart for sequential processing to create new file

and used again in the next run, the dotted line also shows that the tape reenters at that point. In Fig. 3-10, the output tape becomes the input tape during the next cycle.

Preparing reports is the basic task of any information system. When numerous reports are prepared from the same data file, they can be flowcharted as shown in Fig. 3-11.

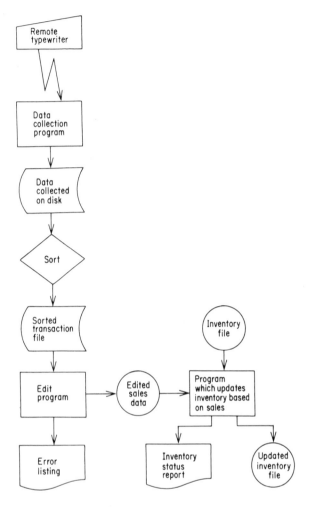

Fig. 3-9   Flowchart for collecting, sorting, and editing data to produce two outputs

TYPE OF FILES

Although there are many ways to store, access and retrieve data, one basic processing method is usually used to keep that data current. This method involves the interaction of a master file and a transaction file.

When logically related data items are organized into a file for permanent use by an organization, the file is usually called the master file. Data created by transactions of the organization is placed into another type of file called a transaction file. The master file and the transaction file are sometimes called the primary and secondary files. The latter are used to update the former. For example, a consumer purchasing goods or services with a credit card creates a transaction. The amount of this transaction must be added to a master file record to update the consumer's account

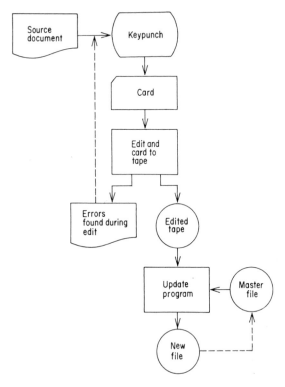

Fig. 3-10 *Flowchart depicting feedback, in which errors are returned to the keypunch section so that source documents may be checked and keypunched over (the broken line from the new file to the master file denotes that during the next cycle the updated file becomes the master file for the next update)*

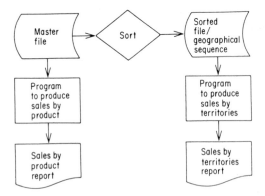

Fig. 3-11 *Flowchart illustrating the production of management reports*

balance. In other words, the purchase of $10 worth of goods means that the balance on the master file must be updated by an addition of $10. Likewise, a payment by the credit card holder to the company means that the master file must be updated by a deduction from the balance. The remaining balance is recorded in a new master file.

Figure 3-10 illustrated a simple update of a master file by a transaction file. Notice that in sequential processing a new master file is created. In other words, the old data is updated by current data and becomes a new file. The same program can be used during the next cycle of updating by simply using its output as the input for the program of the next cycle. This switch is denoted by a dotted line.

In sequential processing, the updating of a master file is accomplished by some key identification. Credit card transactions would probably be identified by the use of a credit card number. Therefore, the master file would be sorted into a credit card number sequence, and so would the transactions. By matching the keys of the master and the transactions, the proper updating of the file can be accomplished. This basic processing which keeps files current is discussed in more detail in Chap. 8. When the systems designer creates the specifications for a system, he must specify each file as a master or a transaction file. If the file is a master file, he must be sure to include a program to update it.

Another type of file occasionally used is called an intermediate or extracted file. When a file is too large to be easily handled for certain types of processing, data can be extracted from it and used to create a condensed file. This file is usually temporary in nature and is sometimes used simply to produce one special report. Systems design specs should allow for extracted and intermediate files whenever they might improve the efficiency of processing.

SYSTEMS FLOWCHART

Figure 3-12 shows a full systems flowchart. This type of system is called an integrated system, one in which several inputs and outputs produce the final desired product.

Figure 3-13 shows a simplified overview of a real-time computer system, one used for hotel-motel reservations. Notice how data is collected through manual input and sent over telecommunications lines to produce reservation forms.

To designate any procedure, one can simply use the written word. For example, a computer program producing inventory control listings can simply be designated as "inventory listing program" or "inventory control listing run." An input-output file should be designated by the name of the file or by some systems code definition. Systems codes are discussed later in this text. Generally, codes are composites of related characters which are used within the system. For example, PAYSYS could be the payroll system. PAYSYS01 could be the first program in the payroll system. PAYSYSAA could be the first data file in the payroll system. For a beginning systems flowchart, that is, the flowchart drawn during the systems analysis phase, it is probably better to use complete words rather than abbreviations.

FLOWCHARTING NONCOMPUTER PROCEDURES

In systems which are not necessarily information-producing but do produce some type of output, flowcharting can be of much help in depicting how the system works. For example, the systems analyst might be called upon to help determine the work flow in a typing pool which does not use a computer but employs various machines. With some modifications, such flowcharts can give new insights into how the work is being carried out. Assembly line procedures, transportation systems, and

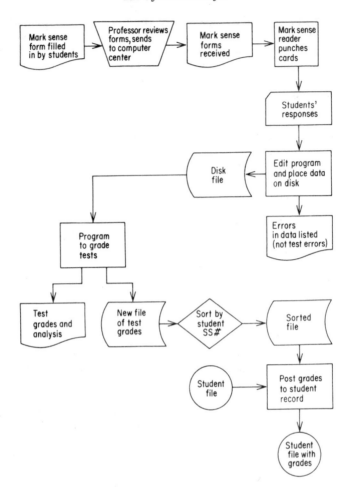

Fig. 3-12 Flowchart illustrating input/processing/output of a system to grade student tests and post grades to student record

the like can also be illuminated by flowcharting; however, other special symbols may have to be used rather than computer system flowchart symbols.

ACCEPTED METHODS OF FLOWCHARTING

The first rule of flowcharting is to fit the flowchart to the person who will be reading it, analyzing it, and using it. The exception to this rule would be the simple case of an analyst using the flowchart simply to develop his logic and methods of processing. However, it should be remembered that the flowchart is primarily a means of communication between the analyst and the users and managers of the system. Therefore, the analyst should avoid technical terms and difficult designations of procedure and logic where standard terms can be used. The second rule of flowcharting

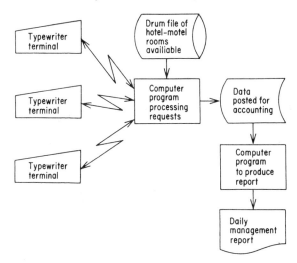

Fig. 3-13 Simplified flowchart of a hotel-motel reservation system (an immediate-response, or real-time, system)

is to use English narrative whenever possible. Avoid the use of systems codes, unless the flowchart is to be used by the systems design team rather than users or managers. Sometimes several flowcharts will have to be prepared to adapt the flowcharts for the different users.

A third rule of flowcharting is to be careful in the use of flowlines. Flowlines normally should be horizontal and vertical to the page or at a 45° angle when depicting input or output from the system. Avoid crossing flowlines when possible. In cases where crossing cannot be avoided. a small hump on the line shows that the lines are not interconnected, as shown in Fig. 3-14.

Parts of the system which are to be emphasized can be set off on the page to draw attention. Plenty of space should be left between symbols, although it is best to use as few pages as possible so that the reader of the flowchart will not have difficulty in keeping track of the logic. Very large sheets of paper can often be used to confine the flowchart to one page. Such sheets are very handy to hang on a wall in the room where the systems team works, providing a means of instant reference.

Another rule of flowchart preparation is to always use standard flowchart symbols and standard names for different devices and files. If a name used in the systems flowchart is used again, it should always refer to the same thing.

Systems flowcharting is not difficult for even the most complex systems if the analyst understands the basic uses of equipment and procedures which are covered later in this book, as well as the basic processing methods such as updating, sorting, report preparation, etc.

The final judge of any flowchart is the person who must use it and understand it. Before presenting any flowchart to management or users, agreement should be obtained from all members of the team that the flowchart does depict simply and accurately how the system will work, although, during the systems design phase, it is possible that new considerations will determine that the flowchart must be changed.

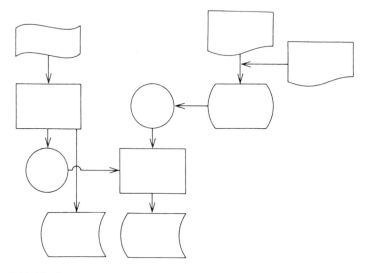

*Fig. 3-14 Flowlines should never cross, if possible; if they must cross, a small hump
should be used to indicate that they are not interconnected; flowlines that join
indicate that a process merges at this point*

A SHORT WORD ON DECISION TABLES

An alternate method of depicting the logic of an individual program, and one
that can be used in conjunction with flowcharts, is the decision table. Decision tables
are probably not used as much as they should be; however, they are not difficult, can be
developed with ease, and do supply some type of easy proof that certain steps should be
carried out.

The systems analyst should be familiar with decision tables and use them when
he feels they are appropriate. Decision tables consist of two parts—the conditional
part and the action part. The conditions are listed before the actions, and the actions
are carried out only if the conditions are met. In the decision table shown in Fig. 3-15,
the conditions are listed at the top and the actions at the bottom. Possible conditions
which can be listed on a decision table are anything which the program or system might
depend upon. Actions are any procedures which should be accomplished either
physically or by the computer system in accordance with the conditions.

An N can be used to designate "No, the conditions have not been met," and a Y
can be used to designate "Yes, the condition is met." Actions can depend upon either
answer. For example, the following conditons might be presented for the handling of a
flowchart:

1. Includes all input-output operations
2. Includes all processing
3. Easy to read
4. Arrows on all flowlines

A line is drawn and then the actions can be presented as:

	1	2	3	4	5	6	7	8
1. Lifetime reserve days contained in bill records	Y	Y	Y	Y	Y	Y	N	N
2. Full days exhausted	Y	Y	Y	Y	Y	N		
3. Coinsurance days exhausted	Y	Y	Y	Y	N			
4. Lifetime reserve days in bill record greater than entitlement*	Y	Y	N	N				
5. Will this bill exhaust lifetime reserve days?			Y	N				
6. Have lifetime days been exhausted from utilization record?							Y	N
7. Guarantee of payment applies	Y	N						
8. Flag as an error—"excess days"		X						
9. Flag as an error—"misapplied days"	X							
10. Adjust lifetime reserve days in the bill record to equal maximum lifetime reserve days allowable (entitled)	X	X						
11. Deduct underutilized "full days" from lifetime reserve days contained in bill record** and add to "full days" in bill record						X		
12. Go to Table A2 ("Days Utilized—Full Days")						X		
13. Deduct underutilized "coinsurance days" from lifetime reserve days contained in bill record** and add to "coinsurance days" contained in bill record					X			
14. Go to Table A3 ("Days Utilized—Coinsurance")					X			
15. Set flag "Lifetime Days Exhausted"		X	X				X	
16. Go to Table A5 ("Guarantee of Payment")	X	X	X	X			X	X

* Entitlement is determined from beneficiary's utilization record.
** Not to exceed "lifetime reserve days" in bill.

Fig. 3-15 A decision table (Courtesy, Social Security Administration)

1. Present to management
2. Redesign or work more

USE OF NARRATIVE

Each flowchart prepared should be accompanied by a narrative which explains in clear and simple English what the flowchart is depicting. Although the narrative may be infrequently used, it might prove helpful to have some of the phases of processing in such a form. The processing parts of the flowchart can be numbered and the narrative numbered to fit.

SUMMARY

Systems flowcharting is the method used to describe a system in diagram form and is the means of communicating the operations of the system to another person or to confirm beliefs about the system to the analyst himself. The flowchart traces the flow of data and methods of producing information within a system. Systems flowcharting can also be used to depict other types of systems other than information-gathering ones; however, specialized symbols may be needed.

Flowcharting symbols are uniformly maintained within the computer industry. The American National Standards Institute and the International Organization for Standardization have established standard symbols and prescriptions for their use and also periodically update them.

An analyst prepares a flowchart based upon his knowledge of the equipment needed at a particular time and upon his knowledge of methods of processing which must take place in a system. Flowcharts may be changed as a system develops; however, early in systems analysis stages, flowcharts should be used to keep communications open between the systems study team.

There are basically two types of flowcharts—the systems flowchart and the program flowchart. This chapter has covered systems flowcharting and shown examples. The program flowchart is covered in another chapter.

Decision tables can also be used to depict the logic of a system; however, they are usually used with program flowcharts rather than systems flowcharts.

Questions

1. Name two reasons for using a systems flowchart.
2. When does the systems analyst prepare the first formal systems flowchart?
3. Name the two types of flowcharts and explain the use of each one.
4. Comment on the following statement: A systems flowchart should show every detail and step to be carried out in a system.
5. Explain the difference between on-line and off-line and how this is handled on a systems flowchart.
6. Input, processing, output has been described as the "sandwich technique." Explain what this means.
7. Name some basic operations carried out by almost all computer systems.

Exercises

1. Using the appropriate symbols, draw a systems flowchart which depicts the following:

THE POST-ENTITLEMENT DATA PROCESSING SYSTEM FLOWCHART

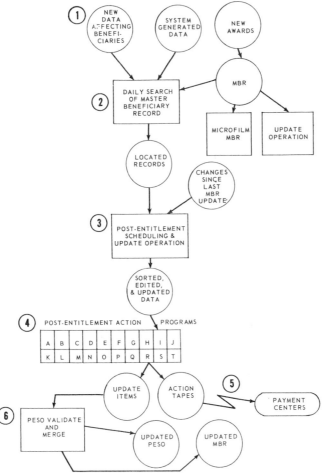

Fig. 3-16 Systems flowchart in use at the Social Security Administration (Courtesy, Social Security Administration)

(a)   A computer program with punch-card input and output of printer listing and magnetic tape.

(b)   The magnetic tape mentioned is used to update a disk file. The output is a new updated disk file.

(c)   The printer listing mentioned is photocopied in triplicate. The photocopies are labeled by a clerk and sent to the company mail room.

(d)   The punch card mentioned is prepared from a sales ticket, using keypunch machines.

2. A magnetic tape device which is not connected to any central processor sends data over telephone lines to another magnetic tape device in a remote location which is not connected to any central processing device. The tape reel at the

remote location is manually removed and placed upon another tape device which is connected to the central processing unit. This tape is used to update a master file and produce an updated file; however, before this tape from the remote location can be used, it must be sorted into account number sequence. The updated file is used to produce a daily management status report. Flowchart all of the above procedures.

3. Analyze the flowchart describing the Social Security subsystem shown in Fig. 3-16. Write a narrative explaining what this flowchart is illustrating.

4. Given the following minimum information:

Salesmen's daily reports are keypunched and the punch cards are used to update the monthly sales tape file. The updated monthly sales tape file is used to update the inventory control file which is located on a disk. The updated inventory control file on the disk is used to update the production scheduling tape file. All three of the updated files—the monthly sales file, the inventory control file, and the production scheduling file—are used in the accounting system and update the general ledger tape file.

draw a systems flowchart—without being concerned as to exactly how all of these procedures above take place—which not only shows all the above procedures, but how they are interrelated, that is, what outputs become which inputs. Be creative in your design and attempt to determine where sorts would be needed and where management reports could be produced.

4

Systems Design

Systems design is the second major phase of system development. After the system has been studied and flowcharts of its probable operation have been developed, approval should be obtained from both the users and the managers of the system. Systems design can then proceed.

Systems design is essentially a synthesis of many techniques taken from the fields of computer science, business administration, engineering, mathematics, economics, and other areas.

The objective of this chapter is to present a general description of the systems design process. The next six chapters present the various tasks and considerations needed for a detailed design. Systems design can be defined as the process of determining exactly how a system should operate and the requirements for its operation, specifying the input needed and the output that is to be produced, specifying what must be done by the computer and also manually to meet these requirements, determining the best way to perform these computer runs and manual tasks, and establishing required performance levels for all the above. In other words, systems dèsign states what the characteristics of a system will be. Systems design specifications can be thought of as similar to the blueprints used by engineers to construct an object.

Although systems design is subject to change as specifications are developed, the final specifications must always explain how the system will work in exact detail.

System designers are often the same personnel who develop the systems analysis and flowcharting discussed in previous chapters. At other times, specialized personnel are used. In either case, the group of personnel responsible for systems design should be well-rounded.

A systems designer should possess the following attributes:

1. A creative mind
2. A knowledge of management needs
3. An understanding of the capabilities of equipment and programming languages
4. An awareness of constraints such as time and money
5. A command of systems analysis and design techniques

As stated before, systems analysis and design is probably both an art and a science. The systems designer must be creative in his designs; however, he must also be attentive to standardized techniques and methods. Only by understanding in detail the objectives of management can he produce a system which will meet with management's approval.

To produce an acceptable system, the systems designer must understand how equipment operates and how computers can be used to carry out the tasks necessary to meet the objectives of the system. There must be a close attention to detail in defining input records, output reports, and the types of control and feedback needed to operate the system. Although systems design can never be thought of as a step-by-step procedure and there is lack of agreement as to how designing should be approached, it is still true that there are many techniques which can be applied by the designer.

One of the first tasks that must be accomplished is to define output in detail. The output of a system essentially consists of the objectives which have been stated by management. Once the output has been determined, procedures to produce it can be developed. The necessary input can be determined; the necessary controls upon this input can be established to insure that the system will carry out its functions; and finally, feedback channels can be specified so that the system can be adjusted to meet its objectives. Throughout the design phase, attention to the interrelation of the system to other systems is most important; for example, how the accounting system will interrelate with production control systems as well as how each subsystem will fit into the overall system.

SYSTEMS DESIGN AND COMPUTER APPLICATIONS

When designing computer systems, it is helpful to think of the five basic components of all systems, that is, input, control, processing, feedback, and output (these tasks are shown again in Fig. 4-1). In other words, design personnel should ask themselves: What is the output of this particular system? What inputs are needed to produce this output? How can the system be controlled? What processing is necessary? What feedback method can be utilized? The design of each subsystem, whether it is manual or computerized, must satisfy the functioning of each of the five parts of the system. The specifications produced by the design personnel will later be turned over to programmers so that the system can actually be programmed and begin to operate.

It should be noted that the systems analyst often has very little defined for him.

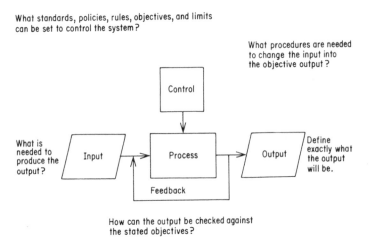

What standards, policies, rules, objectives, and limits can be set to control the system?

What procedures are needed to change the input into the objective output?

What is needed to produce the output?

Define exactly what the output will be.

How can the output be checked against the stated objectives?

Fig. 4-1   Relating the design of a system to the five parts of the system

He has written objectives from management and should know roughly his cost and time constraints. However, no input, output formats, or procedures have been determined, and this is the job of the designer.

TYPES OF COMPUTER USE

There are basically two types of computer use. In this book they will be called (1) integrated systems and (2) immediate response systems. The integrated system is sometimes called by other names such as the stand-alone system, centralized system, batch system, processing system, EDP system, etc. The immediate response system is usually called a real-time system. Essentially, an immediate response system contains a number of intricate control programs, which send and receive data to and from remote terminals connected to the system. After these control programs have been designed, the design of an immediate response or real-time system is much like that of an integrated system. In addition to the computer procedures which must be specified for the two types of system, the designer must consider the manual methods which will be used, such as collection of input data, preparation of data, distribution of output, use of forms, and the specific tasks of the personnel who will be working within the system.

TYPES OF DESIGN

The most common problem faced by the systems analyst is to design a system that will replace an old system. He may be required to utilize existing equipment and therefore must know its capabilities and how it operates. Essentially his task in such a case is to optimize the present system, possibly by recommending new equipment and most probably by recommending new procedures and techniques.

This type of design might be, for example, an accounts receivable system which will fit into a larger accounting system within an organization.

The other type of design which the designer may find himself faced with is that of a new system, or what is sometimes called "total design." In this case, the designer will have complete freedom to design the system insofar as he observes the restraints presented by the objectives of management. It will now be his job not only to design the system but to recommend what equipment should be selected and how it will operate (Fig. 4-2).

Regardless of the type of design, it is important for the analyst to optimize the system in the sense of insuring accuracy, producing more at less cost, and other objectives which are inherent in systems design but not necessarily stated by management.

DESIGN OF FUNCTION

One of a designer's first considerations is what functions will be necessary within the system. The functions might be listed as follows:

1. Creating
2. Receiving
3. Recording
4. Classifying
5. Editing

How will the data be collected?

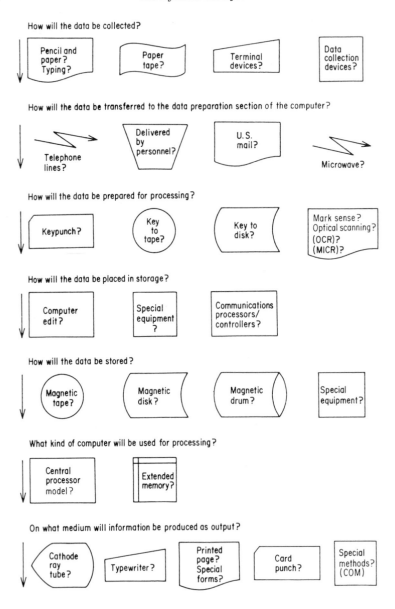

Fig. 4-2   Equipment considerations of the systems analyst

6. Sorting
7. Processing
8. Calculating
9. Summarizing
10. Storage

11. Reporting
12. Distribution
13. Inquiring
14. Auditing
15. Controlling
16. Feedback
17. Manual operations
18. Filing

There may also be other functions which the analyst must consider, but these are the basic ones which are inherent in most systems and will therefore be discussed in detail.

Creating: Data which enters an information system has to have been created somewhere, either outside an organization or inside it. Examples of creating data are transactions such as sales orders, shipping orders, hours worked by personnel, and any process within an organization to which some specific value can be attached. For example, when a worker works 40 hours on an assembly line, he has created the data of "number of hours worked." This number of hours worked must be processed to produce within a reasonable amount of time a payroll check for the worker. The systems designer is very concerned about the creation of data: how is comes about, how it can be gathered at the source, and how it moves from the source into the system which he is designing. In many cases, specialized equipment must be considered for data collection. The flowchart which the designer presents as the final design flowchart should probably begin with the initial creation of data.

Receiving: Receiving of input into a system is another major consideration of the analyst, that is, how data flows from its created point to the point where it is received by computer center personnel. Log books, specifications, and standards are often necessary to control this receiving process. Procedures which workers carry out during the receiving of data are usually specified by the systems designer.

Recording: How data is to be recorded so that it can be used by machines is another decision which the systems designer must make. In today's computer industry there are many ways to record data for machine use. The most common method is still key punching, but with the introduction of optical scanning and data recording with magnetic tape, the decision of which method to use has become more complex and must be considered in detail.

Classifying: Once data is received it must be classified in terms of type, the file it will be stored in, its security, priority, and so forth. The analyst must also consider what codes will be used for classification.

Editing: Items of data coming into the system are edited in many ways, some to insure accuracy, others to reduce the amount of characters needed for recording, and still others to expand the information presented. During output procedures, data is usually edited with such characters as dollar signs, commas, periods, blanks, or whatever is needed to present it in the most useful form. The systems designer must specify the editing procedures required for both the input of data and the output of information.

Sorting: In most systems, data is placed into files in a particular sequence. This sequence is based upon a key identification number such as a social security number, invoice number, customer number, or the like. The sorting process takes place as a program run. The designer of the system specifies when sorts will be made, how they will be made, the keys which will be used, and other such details. It is important for data to be filed in the most efficient manner. For example, it would be most difficult to produce a report showing salesmens' totals by territory if the salesmens' records were in order by their social security number rather than by territory.

Processing: Processing consists of all arithmetic calculations and manipulations of data, such as lengthening or shortening, and especially logical comparisons and other decision checks. Processing takes place in accordance with one or more computer programs. Processing steps are divided up into logical computer runs in the particular sequence needed for execution. The sequence of processing is stated by the systems designer in his systems specifications, usually as a flowchart.

Calculating: Calculations upon data consist of all operations such as addition, multiplication, division, subtraction, etc. What formulas will be used and all mathematical steps that must be taken must be stipulated by the analyst. For example, in an accounting system, debits and credits are essentially additions and subtractions and their relationship to the account must be carefully considered.

Summarizing: Data is frequently summarized both to reduce the amount of storage space needed and to make the data more easily usable as output information. Summaries are usually based upon control keys, such as customer numbers, salesmens' numbers, etc. For example, a report on all items in inventory would be more useful to management if summarized by particular warehouses, product areas, turnover, usage, or some other classification. The summaries to be produced depend upon the users' requirements. The designer must specify where summarizations are necessary and how they will be produced, both to reduce the amount of data needed and to produce the output required by the user.

Storage: To store data into files, the designer must decide on methods of storage, amount of storage, format of storage, etc. All such decisions profoundly affect the efficiency of the system, especially the amount of time required to produce output.

Reporting: Basic output in most information systems consists of reports. The types of reports which will be produced depend upon the needs of management and the requirements of the users of the system. Reports are designed on "report lay-out sheets" as described in another chapter, and these sheets are interchanged between the users and the design personnel until they are approved. Reports fall into several different categories. Some reports are mandatory, such as government reports. Others are useful for reference, such as inventory listings. Others help control the system and aid in decision making. One of the basic types of reports is the exception report, which shows exceptional or unusual conditions within a system.

Distribution: The output of a system is useful only if it is distributed in a timely and efficient manner. Distribution techniques are another item that must be specified by the systems designer.

Inquiring: Once data is stored in a computer system, management and users may make inquiries about the status of a particular portion of the system or request specific reports from the system. Such inquiries are specialized and usually not built into the system's design. However, a decision must be made whether to have a generalized reporting system by means of which inquiries can be made through specialized codes and special reports produced. If the decision is negative, inquiries can be produced by programs written especially for that purpose by a programmer. Numerous inquiries for specialized reports indicate to the analyst that a system is not producing all the data which was really required by the users. Nevertheless, inquiries should not be discouraged since they are sometimes some of the most useful output of the system.

Auditing: Systems should be constructed so that their flow of information and procedures and their input-output can be audited. This is especially true for systems for accounting applications but applies to all types of information systems. In many large organizations, a special auditing section is set up solely to audit and analyze the operation of the system. An audit trail, which is the documentation of the flow of data from its source to its end use, should be designed into the system by the designer and should be available to the Certified Professional Accountants as well as the internal auditor team. Managers should receive periodic audit reports which indicate the status of the system.

Controlling: Systems will not function without adequate control. The various aspects of control are discussed in detail in another chapter, but it may be said here that as processing procedures and methods are developed, control over them must be established.

Feedback: Before control can function, there must be some type of feedback upon which it is based. A common example of feedback is complaints from customers who have been billed erroneously by a computer system. To be able to handle feedback is a major concern of the systems designer.

Manual operations: Manual operations necessary to a system are usually specified by the systems designer and may include simple tasks of nonsupervisory workers or may also concern deadlines and time specifications which must be met by managers and users of the system. For example, determining the transfer of time cards from the foreman to the data processing section is a specification of a manual task that must be accomplished at a certain time.

Filing: Filing of information manually as well as by computer storage media is another job that must be specified by the systems designer. Items commonly filed within a system are log books of computer runs and other such controls of the system.

GENERAL APPROACH TO SYSTEMS DESIGN

Before the details of a system are specified, some general outline of the system should be produced. This outline should consist of a list of all desired output, a list of all required input, a statement of the time constraints and cycles of processing (such as daily, weekly, or monthly), a list of procedures which can be used to produce the output from the input, and a general list of equipment, supplies, and personnel. In other words, the resources which the system will use must be specified in detail.

As stated before, it is difficult to describe a step-by-step procedure for designing a system. However, some general steps which should be taken at the beginning may be listed as follows:

1. Define the problem and state the objectives of the system. This step is usually taken during the systems analysis phase.
2. Understand the present system before proceeding to the next design.
3. Define the subsystems or tasks which must be accomplished within the system.
4. State what each subsystem should accomplish and check each of these tasks against the objectives.
5. Examine the interrelationships between the system and other systems and between the subsystems.

6. Prepare a timetable for accomplishment of analysis, design, and implementation.
7. Determine what the input, output, processing, control, and feedback of each subsystem will be.
8. Prepare documentation for management approval.
9. Obtain management approval or repeat this general design cycle until management and users are satisfied.

The following is a list of considerations which the designer should think about as systems design proceeds:

1. Define each component of the system, such as input, output, processing, control, and feedback.
2. Consider the monitors of the system, that is, what particular type of feedback can be obtained.
3. Provide for ease of feedback. For example, as mentioned previously, when a customer is presented with the wrong billing on an invoice or when orders are lost, etc., there should be an easy method to correct such errors. This feedback should also be analyzed to determine what might go wrong with the system.
4. Since the system should provide for as many exceptions as possible, data codes should be checked, and other such checking methods, which will be described in detail later, should be provided for.
5. The system should be as accurate and error-free as possible. Through the use of input and output control techniques, most systems can be designed to be practically error-free.
6. The systems designer must be cost conscious, and the system should spread costs as widely as possible. In other words, the more use which can be obtained from an expensive piece of equipment or from a program, the less the unit cost will be.
7. The system should be designed and planned with the entire organization in mind.
8. Management needs should be considered at all times.
9. The system should fit into the overall total system of the organization.

LIMITATIONS OF THE SYSTEM

The limitations, or what is sometimes called the boundaries, of the system should be analyzed before detailed systems design begins. Usually, limitations are set by output requirements. Other limitations are imposed by government reporting dates, lack of money, time requirements, and marketing relations aspects of the system such as filling customer orders within a certain time limit.

What the system should accomplish must be analyzed in detail. In many cases, systems have been designed to try to do much more than was realistically possible. By a good definition of objectives, however, this particular problem can be overcome.

NEED FOR STANDARDS IN DESIGN

Before detailed systems design begins, standards for design should be specified in writing. These standards include codes which will be used for data, particular types of procedures which may be used to process data, names of program runs, file names, data element names, as well as general industry standards. It is very important for a systems designer to have a basis for communications. For example, if he refers to item M6493, the meaning of that particular code should be clear to all other systems team members. Many times the analyst will determine the standards during the systems design phase. A company's policies, regulations, and standard operating procedures will also help determine many of the necessary standards.

DEFINING THE SYSTEM

Basically, a computer system requires the determination of the following: (1) design of input, (2) file maintenance procedures, (3) transactions processing, (4) report producing, and (5) answering inquiries. Each of these general categories must be detailed in written systems specifications. There are many steps which the designer must go through to produce such specifications. They may be listed as follows:

1. Management approval is obtained after the systems study is completed to proceed to systems design. Users of the system are contacted at this time for their approval also.
2. Meetings of the systems study group with the systems design group should take place.
3. A detailed analysis of problems and objectives is made, based upon systems study documentation.
4. Detailed studies are made of the systems documents collected.
5. The general systems flowchart is analyzed for possible improvements and changes, and for final acceptance by the systems design team as the basis for their design phase.
6. All other improvements and changes which may have been considered in the systems analysis phase are now discussed.
7. Systems code names, data names, file names, etc., are assigned when possible. These will be subject to some change as systems design takes place; however, there should be a list of codes at the beginning of systems design.
8. Design of outputs.
9. Design of files, data bases, storage mediums, etc.
10. Design of input data collection forms and special forms.
11. Application of control procedures.
12. Program runs defined.
13. Program logic flowcharts developed.
14. Output and input crosschecked.
15. Specification of job control card standards.
16. Specification of programming standards.
17. Specification and summarization of control procedures, feedback methods, processing formulas, techniques, etc.
18. Detail listing of all data elements and their characteristics.
19. Preparation of management narrative.
20. Preparation of systems specifications package.
21. Obtaining of project managers' approval.
22. Presentation to top management and users for their approval.

Obtaining management approval to proceed: Before any systems design is carried out, management and users of the system must approve the analysis phase of the project. At this point, the decision made is probably based upon a feasibility study produced as a subtask of systems analysis, whereas managers and users analyze recommendations from the systems group to decide whether the system should be implemented. If the decision is favorable, management will produce a letter stating the detailed objectives and requirements of the system. This written specification is the basis upon which the systems design is carried out. Management should always be careful to point out limitations such as time and money.

Assignment and meetings of the systems groups: Once the "go-ahead" decision is reached, there should be a meeting between the systems analysis group and the systems design group. (In some cases, the same personnel who did the analysis will also design.) The information obtained from the systems analysis phase is discussed with the systems design group so that a variety of points can be presented and a general discussion can take place as to how the systems design should proceed.

Detailed analysis of problems and objectives based upon systems study documentation: The systems analysis group and the systems design group now make a detailed analysis of all material gathered by the systems study. This documentation, which includes lists of all equipment in the system, rough flowcharts, results of interviews, observations of present systems, etc., is compared to management's statement of the problems and objectives of the system.

Detailed study of systems documents: At this time, some of the documents of the system can be assigned for study to various members of the systems design group. Recommendations for changes and improvement can be made after each member has studied his particular part of the system.

Analysis of the general systems flowchart: The general systems flowchart produced by the systems analysis group is studied by the systems design group. Improvement, changes, and general acceptance by all members of the group are needed at this time. In some cases, it may not be possible to produce the final systems flowchart because of changes which will be made at a later date. However, before the specification of files and processing steps, a general systems flowchart must be developed as a guide to all members of the group. This flowchart is often drawn on a large scale and placed upon a wall close to the working area of the group so that it can be referred to at any time by any member of the group.

Discussion of possible improvements or changes: If the systems design personnel and the systems analysis personnel are not the same, then there should be ample time given for the former to analyze the work of the latter. All systems design personnel should be in agreement with the conclusions and recommendations made by the systems analysis team. If they are not, discussion of the areas of disagreement should continue until all personnel are satisfied that the systems methodology to be used is the most efficient way to meet the objectives of the organization.

Assignment of systems codes and standards: All codes which will be used in the system must be stated in writing and a copy given to each member of the group. As new names are produced or new codes are developed, they should be distributed to all members of the systems design group.

Design of output: The objectives of the systems output is specified by the systems design group at this time. By the use of output layout forms and various output techniques described in a later chapter, the reports and other outputs of the system are specified in detail. These output forms are then discussed with the managers and users of the system. Approval of these output forms should take place before systems design proceeds. In other words, a sample rough form of the output reports is presented to the users of the report. If the users decide that it is in the required form, then the output can be approved and the design for producing this output can proceed.

Design of files, data bases, etc.:   This particular phase of systems design is covered in the next chapter. The designers must determine how data will be formatted within the computer storage media, how it will be used, in what order, what characteristics it will have, and what storage method should be used.

Design of input data, collection forms, specialized forms: The functions of creating data, receiving it, recording it, and using it as input are considered at this time. Basically, the output that must be produced determines the input that will be needed. Use of specialized forms and techniques for input are covered in a later chapter.

Application of control procedures: As processing procedures are developed, the controls needed for these procedures can be specified. The designer specifies what control edits will be made upon the input, what physical and manual controls will take place when data is transferred within the organization, what control techniques will be used on both output and feedback of input data, and so on. Since these controls are not easily determined, they are usually developed over the entire design phase of systems development. New control procedures are added constantly once the system has been implemented and more is learned about its various problems. It is important that they be in writing and impressed upon the programmers who write the programs for the system.

Program runs defined: By means of flowcharts and a listing of the processing needed, the tasks to be accomplished are divided into logical computer program runs. Each of these runs actually constitute a single computer program. Although any one program may interact with other programs, it basically remains but one processing task. Programs can be assigned names as they are defined and can be analyzed as to where they are needed, why they are needed, whether they can be combined into other runs, and whether they are in a logical sequence.

Program logic flowcharts developed: As a program run is designed, the systems analyst should produce a rough diagram of its logic, a procedure detailed in a later chapter. The programmer can often use this flowchart as a basis for development of his program, although it is likely that many changes will develop before coding begins.

Output and input crosschecked: At some point during systems design, the output produced should be compared against the input. In other words, every output item should be checked to be sure that the input item needed to produce it has actually been specified.

Specification of job control card standards: Although the systems designer may not necessarily have to write the detailed job control cards which will be needed, he should specify the standards on which they will be based. Job control cards are specifications which control the computer during processing, a procedure covered in more detail later.

Specification of programming standards: The designer, with the project managers' approval, should specify the programming standards which must be followed by the programmers. These include standard program and file names, programming languages to be used, commenting and annotation of programs, etc.

Specification and summarization of control procedures, feedback methods, processing formulas, techniques, etc.:   At some point, written specifications must be prepared describing what each control procedure is, how feedback will be obtained, and what formulas and other techniques will be used to process the data. All these should be so laid out that they can be easily referred to by any user or person working in the development of the system.

Detail listing of all data elements and their characteristics: Once the output and input have been defined, a listing of all files and records of data within the system should be produced to serve as a basis for referring to all elements within the system. Persons concerned with the system should be informed of any addition or deletion of elements.

Preparation of the management narrative: When the systems design has been completed, one member of the team, possibly the project manager, should prepare a narrative in nontechnical terms which can be read easily by management or users of the system. The narrative should specify in simple English all the above steps and how they are carried out, the thinking of the analyst and the designers as they developed these steps, and how and what they mean to the users and managers of the system.

Preparation of systems specifications package: Besides the management narrative to be used by nontechnical persons, a systems specifications package should be produced which will be used by the technicians, that is, the programmers and other personnel who will operate the system. These specifications are essentially the above steps in written form. A complete systems specifications documentation package is outlined and presented in another chapter.

Project manager's approval: Once design has ended, it is presented to the project manager who decides whether to send parts of it back to the designers or give his approval. Once his approval has been obtained, the entire systems team or selected members will begin to prepare a presentation to management.

Management's approval: Management, along with the users of the system, are usually presented with the entire system, step-by-step, by the systems team. This includes flowcharts, file layouts, data element lists, and all specifications of the system presented in as nontechnical a manner as possible so that the manager and users can see how the system will operate. At this time, and in some cases even before it, training of the users of the system by the systems team should begin.

When all the above steps have been carried out, the systems design phase is essentially completed. All specifications are now turned over to the programming team. In some cases, the designers of the system may also perform some programming duties. However, normally, the programmers are entirely different personnel who must be briefed about the system and their particular jobs.

PERSONNEL CONSIDERATIONS

In many cases, systems design will create new jobs or eliminate old ones. The systems designer should be familiar with the process of writing job descriptions that define the new jobs and also be certain that the elimination of any job is in the best interest of the organization. If any job is to be eliminated or created or have a change in its job description, it should be reported as soon as possible to management for its approval. Job descriptions are included in the systems documentation package and become a part of the specifications.

DESIGN OF MANUAL METHODS

Manual tasks must be performed in most systems. The analyst should be familiar with office equipment and consider carefully the manual steps involved in his design. The systems designer may often be called upon to perform work studies or time and motion studies and should familiarize himself with the necessary techniques.

However, the creation of a truly manual-machine system such as that used in offices for processing paper work is really the job of another person, who is usually called an administrative designer or an administrative systems designer. The words "systems analysis and design" usually connote the use of computers.

DESIGN FREEZE

At some point during the systems design phase, there is a decision to freeze the design. This essentially means that the design is finished, but it does not mean that no changes can be made. It simply means that any change must be made through a formal request. Formal requests for changing the design after a design freeze should be made to the project manager and should be approved in writing.

AUTOMATED DESIGN

Many methods have been proposed and several systems developed which claim to be able to automate the functions of systems design. Some of these systems are for sale by well-known computer software and hardware companies. Supposedly only by specifying the output needed, the system can be designed automatically and flowcharts and specifications produced. Some of these methods work quite well; in fact, they are very good for checking systems designs. However, there is no automated system which is generally accepted within the computer industry and which has been proven over a long period of time.

Probably, in a few years, automated systems analysis and design will become more generally used throughout the computer industry. After all, what is more natural than systems analysts using the computer to help perform their own jobs.

With more research in this area, it is probable that eventually a system will be developed where managers, users, or systems groups will define the problem, objectives, and output desired. A computer will then prepare the systems design specifications, programs, and documentation. In a way, it might be possible for the designer to design a system that will make his own job obsolete. Many authorities are predicting use of the computer to perform systems design in the near future.

SYSTEMS DESIGN COMPLETION

At the completion of the systems design phase, three basic documents should have been produced: (1) the general management narrative, (2) the systems flowchart, and (3) the systems specifications, which consist of various documents to be discussed later.

These three documents form the basis of the operation of the system. They should be filed in a readily accessible place and copies of them given to all persons concerned, except for any confidential material. Throughout the life of the system, these three documents will provide a basis for discussion, analysis, changes, and improvements. Somewhere in the standards book of the organization, there should also be rules as to how to obtain approval for changes in the specifications of the system.

SIMULATION OF THE SYSTEM

To test the operation of the system effectively without actually implementing it is the objective of simulation. Simulation simply means building an operating model of the system and inputting various data and other specifications or constraints to see how it will operate, in other words, to see if it will produce the output desired. Almost any system can be simulated; however, the two great factors which usually prevent simulation are time and money. The building of a model can be a very time-consuming project, as well as very costly.

CONVERSION TO A NEW SYSTEM

As systems development is completed and implementation begins, the management problem arises of converting to the new system. There are three methods used for converting: (1) parallel conversion, (2) phase conversion, and (3) complete conversion. Although the safest is that of parallel conversion, a combination of any of the three may be the best solution.

In parallel conversion, the new system is put into operation at the same time that the old system continues to be used. Results or outputs of the new system are compared with those of the old, and any differences are analyzed to decide whether there might be systems design or programming problems. After the new system has been thoroughly checked out, the old system can be discontinued.

Phase conversion is a combination of both complete conversion and parallel conversion in which some aspects of the system are completely converted while others are run parallel to the old system.

In complete conversion, the old system is discontinued and the new system begun immediately. This is probably the most risky of all methods because there is no assurance that the new system will adequately carry out the objectives of the old system. Complete conversion is usually avoided. Ordinarily, parallel conversion is used to make the transfer to the new system.

SUMMARY

Systems design essentially consists of the development of the specifications which will be used to implement and operate a system. It is both a creative process and a process that follows standardized rules and techniques which have been proven over years of experience. Basically, systems design involves the design of the five elements of a system, that is, input, output, control, processing, and feedback. The designer also pays particular attention to processing techniques and file storage, as well as how to manage data and information.

The systems designer, working from the systems study prepared by the systems analyst, is concerned with designing the functions of the system, such as creating and receiving, classifying and editing, processing and reporting, and distributing data. A general approach to design is to break a system up into subsystems and study each one by examining its elements.

The design, when completed, should be the most efficient method of accomplishing particular objectives. The systems designer must rely on standards for his design, and the project manager must be sure that all members of the systems team have the same assumptions about the problems and objectives of the system.

Important to systems design is the use of written specifications at all times. The system must always reflect the desired needs and objectives of management. An overall view by all members of the team of the system, subsystems, and components is desirable and necessary. The systems designer must leave room for expansion and flexibility. He should explore all alternatives as to how the system can produce the objectives. The system should be continuous in operation; there should be a minimum of human intervention, especially if a computer is used for implementation. System design does not stop with the initial design specifications, programming, and the installation and operation of the system; it also plans for the maintenance, changes, updating, and error corrections which will take place after the operation begins.

The systems analyst, designer, and project manager must plan for the training of personnel who will be using the system throughout the systems design phase.

In summary, the systems design specifies in rigorous form the details of the system. It is both general in nature, showing a complete picture of the system in one general flowchart, and also specific, in that each detail is described in some written specification. After the design phase is completed, the written specifications are filed for permanent reference throughout the life of the system. Written specifications are also turned over to the programmers for use in programming and implementing the system.

As a final step, the systems specifications are presented to management, usually in a formal presentation. At this time, the designer explains his thinking and philosophy and answers questions about the system.

Although the systems specifications are turned over to the programmers, the designers must be available throughout the programming, testing, and conversion phases of the system. As the systems tests begin, changes in design may have to be made. Since programmers usually are concerned with only one or two programs within the system, they do not have an overall picture of how the system operates. They must refer to the systems designer and systems analyst for information as to how to proceed with solving particular problems. Severe enough problems can lead to a complete systems redesign.

Systems design can be defined as the process of determining exactly how a system should operate, determining the requirements for this operation, specifying the input needed and the output that is to be produced, specifying the computer and manual steps which must be accomplished, determining the best way to perform these computer and manual tasks, and establishing required performance levels for all of the above.

Questions

1. If systems analysis is the investigation of a system, then what is systems design?
2. How does the theory that a system consists of five basic parts, such as outlined in the first chapter, relate to the design of the system presented in this chapter?

OUTPUT DESIRED

Packing Slip for Customer #120463

Item No.	Qty.	Description
AA4573	4	Door moulds
AB6175	16	Screwdrivers
AC7216	8	Screens

INPUT TO BE COLLECTED

Customer number from sales ticket
Item ordered from sales ticket
Quantity ordered from sales ticket
Description from inventory file

SYSTEMS FLOW CHART

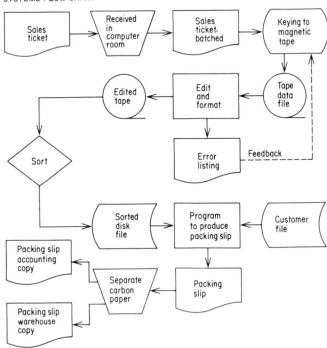

Fig. 4-3   Flowchart for Exercise 2

3. Name four functions the systems designer will probably consider before data ever enters the computer.
4. Categorize the twenty-two steps of systems design listed in this chapter in terms of five basic phases of systems design.

5. Make a list and discuss at least four places in the process of systems design where standards are most important.
6. Discuss whether systems design will ever become completely automated. Make a note to yourself to answer this question again after completing Chap. 13 of this text. See if your opinions have changed at that time.
7. Prepare a generalized list of the material that the systems analysis group turns over to the systems design group. Make another list specifying what the systems design team will turn over to the programmers.

Exercises

1. Based only upon your knowledge of writing checks and receiving statements from your bank, try to determine how each of the 18 functions listed in this chapter such as creating, receiving, recording, classifying, etc., would be incorporated into a bank check-processing system. State how you believe your local bank handles these functions, using a computer. For example:

   Creating—The date and amount of a check are created by the act of writing it. Checks are gathered together by an individual or a business and then deposited in a bank.

2. Imagine that you have been given the information entitled "Packing Slip Subsystem" (Fig. 4-3). Using the 22 steps outlined in this chapter for defining a system, write a report to the systems design team explaining how they should proceed to prepare their specifications. In your report, answer the following questions:
   (a)   How many computer programs will be needed in this subsystem? How many sorts?
   (b)   Assuming that the packing slip will be printed in customer number order and that the items on it will be in inventory item number order, how would you sort the file? What order should the customer order file on the disk be in?
   (c)   Assume that the warehouse manager has looked at the output packing slip and requested that the customer's name, address, city, state, and zip code be printed on the packing slip. What changes would this cause in the system, if any?

3. While working on the systems design for the packing slip subsystem, you discover that one clerk and two typists will no longer be needed when the system begins to run on the computer; however, two keypunch operators will probably have to be hired. How will you handle the elimination of some jobs and the creation of others? At what time during the implementation of the subsystem will you recommend that the change take place?

5

Data Management

Data management is the process of organizing data into a logical structure and providing methods of access to this data. There are several subareas of data management such as processing methods, standardization, security, and control. The objective of this chapter is to present the concepts, principles, and techniques which the systems analyst uses to make decisions about the organization, structure, and application of data. Data has been defined previously as any character (alphabetic, numeric, or special) or group of characters. Data is created at many places within the organization and is used to produce required information. However, most data is not considered to be information until it is processed and presented as output.

The use of magnetic tape and disks as data storage mediums means that there are many options available to the systems analyst and designer for data management. Without standard methods and manufacturer specifications to rely on, the analyst and programmers would be confronted with hundreds of different methods of organizing data. Many of these would be illogical; however, the imagination of the systems analyst or designer is the only limit to the way data may be organized. To overcome the lack of standardization and to present principles and methods which have proven themselves over many years of use, each computer manufacturer supplies with his computer operating system a subarea usually called Data Management System or Data Management Subsystem.

The systems group should know the type of computer system which will be used, but whatever the type, the design of data files and methods of processing will be similar. Most data organization methods and concepts are common to most manufacturer's Data Management Systems, and although there is no definite standard within the industry, the computer manufacturers have tried to use methods which have proven to be the best over the years.

Usually, data management methods depend directly upon the computer manufacturer's data management system. Although it is possible for systems designers to develop their own data management system, that is, to create their own methods and programs, the complexity of the task and lack of time and money usually prevent this decision.

A DATA MANAGEMENT SYSTEM

A Data Management System supplied by a computer manufacturer usually consists of three basic parts:

62

1. File creation and organization conventions
2. Input-output access routines
3. File support routines

The file creation and organization conventions are those rules which specify how data may be organized and used. Methods termed sequential, direct, and indexed sequential are the three most common ones of organizing and accessing data and are specified in most manufacturer's systems. Whatever method is used, the rules specified by the manufacturer must be followed and all limitations considered.

The macro routines supplied perform the input-output tasks needed by computer programs. Access routines provide for label checking, reading and writing individual records, positioning of magnetic disk recording heads in the most efficient manner, detecting and correcting errors, and the simultaneous operation of the central processor with the peripheral devices.

File support routines provide utility functions such as loading cards onto disk, dumping the contents of disks to printers, adding and deleting files, listing the names of files on a disk or tape, and preparing new disk packs for initial use.

DATA MANAGEMENT PROCESSING ROUTINES

The function of data storage and retrieval depends upon computer programming instructions. Since these instructions can be very complex, tedious, and time-consuming to code, they are usually supplied by the manufacturer. Such routines are usually called macro routines or prewritten routines. The macro routine allows a statement given by the programmer to be integrated into a program to perform a particular function. For example, if the systems designer wishes a file to be accessed sequentially, the macro routine which performs sequential access will be brought in during program compilation. Likewise, checking of label records is accomplished by routines which have the ability to access the file and check labels against the programmer's specifications. All access to and organization of files is accomplished by manufacturer supplied macro routines unless, of course, the programmer writes his own.

SYSTEMS DESIGN CONSIDERATIONS FOR DATA MANAGEMENT

The systems designer considers data management from the following viewpoints:

1. Structure of data
2. Organization of data
3. Access of data
4. Limitations on methods of processing
5. Standardization of use of data
6. Security of data
7. Data control

Structure of data: The question of how data should be structured depends upon the hierarchy of information. Starting at the bottom of the chart in Fig. 5-1, the lowest possible form of data is seen to be the bit. A bit is any representation which can

DATA	MADE UP OF
DATA BASE	FILES
FILES	RECORDS
RECORDS	ITEMS
ITEMS	CHARACTERS
CHARACTERS	BITS
BITS	10100110010
10100110010	TWO-STATE DEVICES

Fig. 5-1   The hierarchical structure of data

have only two states. A bit may represent 1 or 0, on or off, open or closed, right or left, up or down, etc. One bit, as stated before, is the least amount of data that can be presented. By organizing bits into groups, characters—the next highest level of data structure—can be represented. Any decimal number can be created from the four bits, 8, 4, 2, and 1, which represent the binary positional values. For example, by turning on the 4 bit and the 2 bit, the number 6 is represented. A group of bits can be (1) a binary field, that is a group of binary numbers; (2) a hexadecimal character; (3) a decimal character, and (4) a single character. The addition of more bits produces alphabetic characters, special symbols such as dollar signs, etc.

A character can consist of any symbol which the computer manufacturer decides to place within his system. Of course, the most obvious characters are the numerical digits 0 through 9 and the alphabetical letters A through Z. Other special symbols are also used, such as commas, dollar signs, negative signs, asterisks, and parentheses. The computer manufacturer's decision as to how many characters his particular system will have is based upon the cost of the memory bits needed to represent the characters as well as other cost considerations.

A single character can be considered as data. For example, the character *M* could be a code for marital status. As the bits and characters are built up, data formations begin to appear. The next highest formation of data is called an item (also a field or element).

Items can be broken up into as many subitems as needed. It is the systems designer's job to decide how many characters will be used to form the items, what the item will consist of, etc. Examples of an item are a social security number, which is a group of digits, a name, which is a group of alphabetic characters, and an amount, such as a tax amount or a payroll amount. Groups of items which are logically related make up what is known as a record.

A record consists of all the data items logically related to one particular object, person, or other type of information (the record is called an entity by IBM). For example, all the information about one employee would constitute an employee record. All the information about one inventory item would constitute an inventory record.

A group of related records make up the next higher structure of data which is called a "file." For example, records for all the employees in an organization make up the employee file. All the information for accounting makes up the accounting file. All the information about sales orders makes up the sales order file. Any logical group of records makes up a logical file. A particular file is used only when the information it contains is needed by a computer program. In other words, an employee file is used only when some information or report about employees is needed. This would seem to indicate that there may be one higher level of data, and there is. It is called a data base or data bank.

A data base can be thought of as an integrated group of items which is accessed by many programs. In some organizations the number of data items ordinarily used is limited enough so that all the items can be placed in one large file, which is usually called the data base. Almost all programs within that organization's system would access this one file.

Data, as we have noted, can be structured by levels, from the complete integration of all data in the data base down to the lowest form of data, the bit. The systems designer considers many factors in structuring data. Starting with a single character, he must decide what characters will make up the items. For example, one common problem is how many characters to leave for a person's name. Some people have very long names; other people have very short ones. This type of decision is based upon various considerations, usually called trade-offs. A trade-off involves something being lost at the same time something is gained, and vice versa. For example, there may be a loss of some characters in long names if the item is too short. However, if the item is too long, there will be wasted space in the record and the file, making for less efficient processing. The decision about the kind and number of characters to be used to make an item is one which the systems designer must make carefully, cautiously, and with consultation between the users of the system and other members of the systems team.

Once the items have been decided upon, the designer must indicate which of them will make up a particular record. For example, should the record consist of such items as race, sex, age, height, weight, etc., or should it be limited simply to names, addresses, phone numbers, and perhaps rate of pay? Such decisions can be made only after careful consideration of the objectives of the system and an interchange of ideas with the managers and users of the system.

There is no formula to determine whether files should be kept separate or in one large data base. The decision whether to have a data base depends upon many variables. If a data base is accepted, it is usually assumed that it will serve as the basis for most of the processing of the system. However, as requirements change, whether because of new government regulations or organizational changes, the data base may prove insufficiently flexible to adjust to information processing objectives. Sometimes it is better to keep logically related types of information in separate files. However, by careful analysis of this problem, it is usually possible to create a data base consisting of the most important data items. Less important data can be kept in separate files.

Organization of data: The most common method of organizing data is in some particular sequence based upon key identification numbers such as social security number, invoice number, customer number, inventory number, etc. Usually, records are organized within a file, from the lowest number to the highest. Once data is placed in sequential order, transactions processed against this data are also placed in the same order.

Other common methods of organizing data are known as random and indexed sequential. In the random file there is no fixed order to the placement of a record. It may be located anywhere within the file and is accessed by some identifier.

In indexed sequential files the records are placed in sequential order. An index at

File organization	Transactions to file	Access
Sequential	Sequential	Sequential
Sequential	Random	Random*
Random	Sequential	Random
Random	Random	Random
Indexed sequential	Sequential	Sequential
Indexed sequential	Random	Random

* Most DMS will not allow this combination.

Fig. 5-2   Methods of access and when they should be used

the beginning of the file contains the exact location of each record, thereby allowing direct access to it.

Access of data: The question of how data should be accessed depends, of course, upon how it is organized. It is also very dependent upon how the input used with the file is organized. For example, it would be very time-consuming, if not impossible, to process random input against a file organized sequentially on a magnetic tape. The basic methods of accessing data are known as the sequential, direct, and search method.

Sequential access means that the records within the file are read or written in sequential order, that is, in the same way they are organized. Direct access means that order of access and position of the record in the file have no significance but that the system will access the record straightaway. Search methods of access are methods where an index or tables are used to find the address of the record within the file. Search methods also include binary searches of a file that has been divided up into halves and quarters, etc.

There are basically two considerations when processing files. One is the organization of the transactions, which contains the key used for the retrieval of the data. The other is the organization of the data file itself. Therefore, data processing methods can be thought of as being sequential/sequential, meaning that the input is sequential and the data file is in sequential order, or sequential/random, meaning that the input is in sequential order and the data file is in random order, or random/sequential, random/random, sequential/indexed, sequential/random, etc. See Fig. 5-2 for a summary of the various methods of organization and processing. The decision as to what methods will be used is based upon time restraints, methods of processing, types of keys, and the particular characteristics of the data itself. Generally, it can be stated that if a data item is required immediately, that is, for an immediate response system, such as those used for airline reservations and hotel-motel reservations systems, then there must be some type of direct access to the data. For example, in a chain of 2000 motels, a person requesting a room in the 1999th motel would not be happy about waiting while the system searched sequentially from the first motel to the 1999th motel to make the reservation.

If very little input has to be transacted against a very large file, direct access is usually the best method; however, if almost every record in the file requires a transaction, then the input file and the data file should probably be sequential. In summary, the access method used depends upon the need for immediate response and the volume of transactions which will be processed against data files.

Limitations on methods of processing: When considering methods of processing, the systems designer must take into account not only the organization and access of data, but also the capabilities of the computer system. For example, in some systems there is no direct access capability, and it therefore must be eliminated from consideration. The type of programming language used is also important in determining methods of processing. Some subsets of COBOL may have limited direct access capabilities. In most cases, direct access methods are not used with FORTRAN. Capabilities of these languages are explained in a subsequent chapter.

The instructions which the computer is capable of performing are also a consideration in methods of processing. Every computer has a limited number of instructions which it can carry out. Most computers can add, subtract, multiply, divide, perform input-output, and move data around for manipulation. However, the ability of the computer to perform table look-ups, search routines, and other types of routines which the manufacturer may or may not supply help to determine the methods of processing which will be used.

Standardization of use of data: Every system must have some means of standardizing the use of files and the names which will be given to the data items and records. Standardization of data names, data items, etc. is the decision of the systems designer unless they are specified in a standards manual for systems design. Data management systems supplied by the manufacturer of the computer also influence standardization. For example, label records have standard formats determined by the manufacturer. Although a system may use nonstandard labels, it is usually more efficient to use the standard formats which the manufacturer supplies. These labels on tape or disk are magnetically recorded at the beginning and the end of each file.

Names for particular items of data such as passwords, keys, code numbers, etc., can also be standardized by certain characteristics of the manufacturer's data management system. Before systems design proceeds, the systems team should become familiar with the principles presented in the data management system of the particular computer to be implemented.

The manufacturer's area of standardization also encompasses the methods used to catalog files upon a storage medium such as disk or tape and to retrieve indexes and directories of cataloged data files.

Security of data: Security of data depends on both the systems designer and the manufacturer's data management system. Types of security which are of concern to the systems designer are both physical security and security of access. Physical security is achieved by the use of good library principles (such as cataloging and indexing), safe storage of tapes and disks in a physical library, and the use of standardized labels on the outside of the disk or tape reels for easy readibility by computer personnel.

Security of access can be attained by the use of passwords. The operating system can be programmed to allow access to a data file only if the program accessing the file presents the correct password. Passwords prevent confidential files from being accessed and also prevent multiprogramming and time-sharing systems from accessing the wrong file accidentally.

Data control: The control of data is a major concern of the data management system and of the systems designer. Aspects of control of data are covered in detail elsewhere. Types of control include priority control, where particular priority numbers allow immediate access before lower priority requests, and physical and internal labels, which specify the names of the files so that only the correct name can access them. File labels also contain record counts so that the number of records expected to be in the file can be compared with the actual record count in the trailer label, that is, the label at the end of the file. The beginning and ending of files are also marked by the data management system. For example, the Honeywell computer system uses a 1 HDR as the mark for the beginning of a file and a 1 EOF to indicate its ending. IBM and other systems use comparable "end of file" and "beginning of file" marks to allow a system to determine automatically when the limits of a file have been reached.

In summary, the systems designer faces a major task when trying to decide what type of data management he will use. His decisions will determine how the data is structured, how it is organized and accessed, what methods of processing will be used, what standards will be applied, and what security and control of the data is required.

FILE HANDLING

The systems designer is concerned with the types of processing that will take place within a system. The main types of processing are sorting, merging, updating, and producing output. The designer should produce written specifications for the cycles of this processing. In other words, is it to occur daily, weekly, or monthly? Sorting, merging, and updating are discussed in other parts of this book and will not be expanded upon here. However, the placement of particular computer runs within the system to perform sorting, updating, etc., is an important decision which must be made by the designer, and this decision should be reflected in the systems flowchart. For example, if a transaction is to be processed sequentially against a master file, the designer must think about what sequential order the files are in. If the two files are not in the same sequential order based upon the same key identification number, then a sort will be needed before the processing can occur.

DATA MANAGEMENT SPECIFICATIONS

Before the manufacturer's data management system can be used in a program, the characteristics of the files must be specified. The written specifications needed are:

1. Length type of records
2. Blocking
3. Organization
4. Labels

Length type of records: Records can be organized in terms of fixed, variable, and undefined lengths. Fixed-length records are all of the same size (the same number of characters). Variable-length records vary in length. Variable-length records should be used when large amounts of data are to be stored and record size varies greatly. Employee records would probably be of the fixed-length type because the same amount of information would probably be needed for each employee. However, records of grades of college students would probably be variable because a freshman would have only a few grades to record while a senior or graduate student might have hundreds of grades and recording space would be wasted if freshmen student records were the same size as senior student records. Some data management systems handle variable-length records very well, some are less efficient, and some do not handle them at all.

Undefined records are not common but can be of value when the data transmitted is not defined as to length, such as in the use of telephone communications lines. For undefined records a beginning marker and an ending marker are needed.

Blocking: Records can be considered to be of two types: logical and physical. A logical record is the actual record used within a system. A physical record is a group of logical records. The reason for using physical records is to reduce storage space and speed up input-output.

One of the biggest problems with computer systems is the great difference between processing speed and input-output speed—sometimes referred to as the "I/O bound" problem. In other words, a system is bound to its input-output speed. For example, a card reader can read possibly one card every one one-hundredth of a minute, whereas the central processor can carry out billions of instructions in a minute. The difference between the number of instructions the processor can carry out and the time that it takes for the input record to enter the system is the "I/O bound" problem, that is, the time that the system waits upon the input-output data is simply wasted time. Several methods have been proposed to solve this problem, and one of them is blocking, that is, placement of several logical records into one group of records called a block. Several logical records can be brought in and out of a system at the same time. The number of logical records placed within the physical record block is called the blocking factor. For example, a blocking factor of 10 to 1 would denote 10 logical records within one physical record.

When making decisions on blocking factors there are several trade-offs. The more logical records placed in a physical record, the more memory locations which will be needed within the system to handle these records. To understand what happens within the system, the concept of buffering must also be considered.

When a record is brought into the system (or is ready to be discharged as output from the system), it can be placed in a memory area called a buffer, which is a holding area. In other words, the processing of the computer does not have to wait upon the physical input and output from mechanical devices because the data can be processed from the buffer which is actually in the memory. Double buffering is the concept most used. Figure 5-3 shows double buffering in diagram form. Notice that the input records are in one buffer and are moved to a second buffer as the processing steps use the input data in the first buffer. The device which handles this input-output is usually called a channel. Therefore, there is input-output transfer at the same time there is processing. This concept is called process overlap. It is not important at this time to understand exactly how the computer system works; however, it is important to understand that trade-offs exist when assigning blocking factors during systems design. Blocking

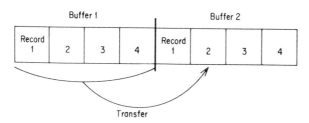

Fig. 5-3   Double buffering

factors mean that more memory locations are required internally. Therefore, large blocking factors might use more memory locations than the computer has available. For example, if a record contains 100 characters and is blocked 10 to 1, then 1000 memory locations will be required in the buffer to contain this physical record. If the system is double buffered, then 2000 memory locations will be needed to contain both buffers. If each data file coming in and out of the system is also double buffered, the same number of locations (for example, three data files would be 3 times 2000, giving 6000 memory locations) would be needed just to handle the input-output buffers. If this amount of memory is available, then the systems designer could assign a 10 to 1 blocking factor. If this amount of memory is not available, he would be forced to assign a smaller blocking factor. Since long physical records brought in and out from disk and tape files are subject to transfer errors, blocking factors are clearly a trade-off between the amount of memory available, the speed of processing, and transfer capabilities.

   Organization: The methods of organizing data have been discussed previously. When determining program runs, job control cards, and other systems specifications, the systems designer must specify whether files will be organized as sequential, indexed sequential, or direct access files. The method of access must obviously be specified for the programmer before he begins to write his programs. All these specifications must be in writing.

   Labels: Labels contain such items as the name of a file, the date of its creation, and the date after which it will no longer be used. These specifications can be stated on control statements and entered into the system during processing. The system will check the information in such labels to control errors in using files and for security of data. The type of label to be used by a system must be specified in writing.

GENERAL APPROACH TO FILE DESIGN

   To design a file, a systems designer must do three basic things:

1. Select the keys of the file
2. Describe each item within the file and decide where it will be placed within the record
3. Select the storage media and access methods for the file.

   Keys: In selecting keys, the systems designer will want to choose identifying items which most closely describe the particular record. Common keys are social security numbers to identify personnel, inventory numbers to identify inventory records, customer numbers to identify customers of the firm, vendor numbers to identify organizations which sell products and services to the firm, and so forth. Countless types of identification numbers exist. For example, credit card holders have a credit card number; bank checking account users have a bank account number. All these can serve as keys for identifying records for processing and retrieval.

   Keys can be major or minor, either the main identifier or secondary identifiers. For example, a major key might be a geographical area, and states within that geographical area could serve as minor keys.

FILE DESIGN

   File design requires the description of the characteristics of the file and the records which make it up, including each item within each record and its position. The considerations which are important in describing the file and the data record are as follows:

1. Life
2. Type of record
3. Actual length
4. Addressing methods
5. Method of storage
6. Technical requirements
7. Representation
8. Future area
9. Symbolic name
10. Limitations
11. Usage
12. Layout
13. Cataloging
14. Security
15. Format
16. Keys

These concerns should be developed into written specifications as described in the following paragraphs.

Life: The life of the data file should be specified in terms of the updating cycle (daily, weekly, or monthly), retention periods, and the date after which the file may no longer be used, if known at the time. An example of such a specification might be: Payroll, updated weekly, not used after accounting closing at end of year, retain for five years.

Type of record: As stated before, records can be fixed, variable, or undefined. The exact type must be specified in the description of the file that the designer prepares. Blocking factors can be stated at this point also.

Actual length: The total number of characters contained in the record should be specified. The length of all data items placed together (usually with no blank spaces between them) make up the length of the record.

Addressing methods: The methods by which the data record will be addressed, that is, random, sequential, or indexed sequential, must be specified.

Method of storage: The method of storage of the record must be specified, whether magnetic tape, disk, card, or any of the many other types of file mediums available.

Technical requirements: Some Data Management Systems require data space in the record to handle certain types of storage and access. An example is the variable length record, in which data space must be left for the IOCS system. The file organization rules should be checked for this requirement.

Representation: Representation in the encoding or compaction of data essentially refers to the form the codes will take and what the codes will represent. Consider the marital status of personnel in a personnel file. There is no reason to have the words "married," "single," "widowed," or "divorced" in the file; instead, a code can represent each status. For example, using S for single, M for married, W for widowed, and D for divorced would be one method. However, in some cases it might be better to use numeric codes; therefore, "single" could be indicated by 1; "married" by 2, "divorced" by 3, and "widowed" by 4. Decisions about representation must be made by the analysts and designers for each item in the file. The codes selected, of course, should be standard throughout the system.

Future area: In some systems it may be determined that more data items will be needed in the future. To avoid having to change an entire file at that time, some blank space can be left which can be defined as "For Future Area Use." In other words, any data items to be added later will have a place already provided for them in this particular part of the record.

Symbolic name creation: Another job of the systems designer is to create symbolic names to describe the data field. Usually, he does so by listing every data item by its actual name or phrase that describes it and then creating a symbolic name which conforms to the rules of data name creation for the particular computer system that will be used. Some systems state that no name can be over six characters in length and must begin with an alphabetic character. Therefore, a data item such as "social security number" could be abbreviated, for example, as SOCSEC. The symbolic name created should be logically related to the data item. In other words, it is more desirable to name a data item for social security numbers SOCSEC than it would be to name it XYM because XYM has no logical relationship to social security. Other examples might be INVNO for inventory number and CUSNO for customer number. Careful attention must be given by the systems designer to this phase of data design. Creating symbolic names for the data items standardizes the names of the items throughout the system and, hopefully, throughout the organization.

Limitations: The limits, parameters, and illogical values of data should be stated. For example, if no inventory items cost more than $200 apiece, this should be stated so that the programmers can insert this control check into their programs. Such control checks are discussed in detail in the chapters on "Input," "Output," and "Control." Limits are usually easy to establish and should be used as much as possible. Illogical values and parameters can then be checked. For example, if the range of a code is established, such as marital status 1, 2, 3, 4, the code should always be within the range of one to four. Any value outside of this range would mean that there is a key-punching error or some other error in the creation of the data.

Usage: The usage of the data items should be described, if possible. For example, for hours worked by an employee, it could be stated that "this data item is used to compute pay during payroll runs." Uses of the data vary from one system to another, but some general statement of how the data will be used should be given.

Layout: Positions of data items in records is not critical in most cases, but it should follow some logical sequence which will be easy to read and understand. A record layout form should be used to describe the items and their location in the record.

Cataloging: Each Data Management System has its own methods of cataloging data sets and files. These cataloging methods should be understood by the systems designer so that he may specify those he feels most desirable in written specifications.

Data files used in data management systems must be given symbolic names. The programs which process the files should also be given names. Data management systems which the manufacturer supplies specify what these names can be. For example, a standard IBM file name must be from one to eight characters in length.

Security: The confidential nature of a data file and the programming methods to control this security, such as passwords and the like, should be stated in writing. Physical security of data is usually stated also. However, physical security is a management function and may be covered in the systems specifications management narrative or when stating control procedures.

Formats: The way the data item will be created, the number of characters in the data item, the decimal points which the numerical fields will contain, and the way the data item is either reduced or expanded, should be stated. For example, a social security number is usually written with two hyphens—one between the third and fourth characters and the other between the fifth and sixth characters. A typical social security number would thus be 430-72-7579. However, in a computer system, there is no need to store the data item in this form. In other words, the hyphens are not needed. Therefore, a social security number should contain only nine numeric digits. The hyphens can be added by the output process if they are needed for output display.

Written specifications such as the above must be made for each data item. The problem with some data items such as names, addresses, cities, states, etc., is the lack of knowledge about the make-up of such items and how they may change. For example, how many characters will be needed for the longest city name in the U.S.A.? The number of characters which should be provided for is sometimes difficult to determine and will be discussed in more detail in Chap. 7.

Selection of keys: Every record should have some means of identification which is unique to that record. For personnel, the social security number is usually the key; for customers, an account number is used. As stated before, these keys are used for identification, processing, matching transactions, and sorting the files.

Keys are often obvious. However, they are sometimes hard to create. For example, when creating a file of grade school students within a large educational system, it would be next to impossible to identify the students' records by social security numbers, since most young people do not obtain a social security number at that age. Such problems can be overcome by assigning arbitrary numbers to the records or by using some other key, such as the first five letters of the last name. Data items selected as either major or minor keys must be specified in writing in the data design specifications.

Carrying out all the above steps will mean that the data files have been designed in general but not necessarily in detail. It is not necessary, however, to define each input and output file at this time. As the data design progresses, detailed specifications of each input file and each output file can be made. However, all subsequent descriptions of input and output files should be based upon the basic data descriptions given here.

SELECTION OF STORAGE MEDIA AND PROCESSING METHODS

As stated previously, the basic methods of storing and accessing data are known as sequential, indexed sequential, and random. The systems designer must make the decision as to how the data will be stored and how it will be accessed. Basic advantages and disadvantages of each of these methods are possible to analyze.

Advantages of sequential processing: One advantage of sequential processing is ease of processing, since the only logic involved is simply to input and output the records one at a time as they appear in the file. Another advantage is efficient storage since records are stored one after another with no blank records between them as may be the case with random files. Control of sequential processing is usually easier since record counts and control totals can be computed as they cannot always be with random files.

Disadvantages of sequential processing: The basic disadvantage of sequential processing is the time required to find any one particular record within the file. If the five hundredth record is desired, it is necessary to process from the first record to the four hundred ninety-ninth before it can be retrieved. When immediate response is needed from a computer system, sequential processing cannot be used. Also, sequential files are inefficient to update because the entire file must be recopied even if only a few records are to be changed.

Advantages of indexed sequential methods: The advantage of index sequential files is that they can be processed both randomly and sequentially. Indexed sequential files, however, use storage space for the index and seem to be best adapted to smaller files. When larger files are to be processed, the space needed for the index becomes quite large, and the sequential method or the random method is probably better.

Disadvantages of index sequential methods: Another disadvantage of the indexed sequential method is that the records may overflow the storage available when updating takes place. Also, the indexed sequential method is not available with all computer systems and not supported by all manufacturers' data management systems. Therefore, the systems designer must be assured that the computer he is using will support indexed sequential methods before specifying them. Indexed sequential record files are slower to access directly than a random file. When immediate response is needed, there will be more time delay with the indexed sequential methods than with the direct or random method.

Advantages of direct access: The biggest advantage of the direct access method is the ability to retrieve a record immediately without processing any other record. Obviously, this allows immediate processing of a request.

Disadvantages of direct access: There are two basic disadvantages to the direct access method. First, storage costs may be greater than those for other methods. For example, sequential files can be stored on magnetic tape at a cost of approximately $20 per reel, whereas direct access files must be stored on disk packs or some similar storage unit at a cost as high as $500 per pack. Second, the direct access method is more complex to use. Keys may have to be transformed by complicated mathematical routines, and program coding may be more difficult.

SUMMARY OF FILE ORGANIZING METHODS

Although most file organization is either sequential or random, it will be worthwhile to summarize all methods commonly used for file organization. These can be listed as follows:

1. Sequential files
2. Random files
3. Indexed sequential files
4. Linear lists
5. Rings
6. Pointers

Sequential files: Any file that is in some type of order based upon a key number is considered to be a sequential file. In other words, all records ordered by an account number would constitute a sequential organization of a file. Sequential files are accessed from beginning to end, one record at a time.

Random files: If the key identification of a record has some relationship to the actual disk address, the file is considered to be a random file. For example, if record number 30 is the thirtieth record in the file, by accessing the thirtieth record directly, it can be said that the file is being directly accessed. There are many ways to establish the relationship of the key number to the disk address. In certain cases the key number will correspond to the cylinder and track numbers of the disk. In other cases, the key number must be transformed by complicated mathematical techniques called "key transformation."

Indexed sequential files: Indexed sequential files are those files which contain an index at the beginning of the file which references the positions of all records in the file. For example, if account number 1006 is located at disk address 2341, then index number 1006 will have 2341 listed in the index. This method allows the system to find the disk address in the index and then move directly to the record in accordance with the address in the index.

Linear lists: Files called linear lists are essentially waiting lines or queues, that is, records come into the system and are placed in waiting lines. How these waiting lines are accessed depends upon the type of list. Two of the most familiar methods are first in, first out and last in, first out. First in, first out (FIFO method) essentially means that the first record to enter the file is the first record to be retrieved. Last in, first out (LIFO method) means that the last record in is the first one used. Linear lists have a beginning record and an ending record.

Rings: A linear list enclosed in an imaginary circle, that is, a list without a beginning or an end, constitutes a ring file organization. Such files are used in advanced processing.

Pointers: A file organized by pointers is one in which each record contains values which tell what record will be retrieved next. These are also sometimes called chain files, in the sense that each record in the file leads through the file to the next record. Pointers also refer to items, that is, each item contains information as to where the next item is located.

In organizing data, combinations of the above methods may be used.

DATA SETS

The term "file" has been used for many years in the computer industry, but a new term is rapidly replacing it, namely data set. A data set is any group of data placed upon some computer recording medium such as magnetic tape or disk.

In modern computers all characters stored on magnetic disks are considered to be stored in data sets. Data sets are of two basic types, partitioned and unpartitioned. The partitioned type includes two subtypes, program data sets and procedure data sets. Data sets which contain data such as master files, transaction files, extracted files, and/or data bases are called unpartitioned. Such data sets are catalogued in the system and can be retrieved by name.

All programs are stored in one large partitioned data set and can be retrieved by name also. Partitioned data sets that store job control cards are called procedure data sets. Data sets such as the type used on an IBM computer can contain all the information needed by a system for processing.

CODES

Three different types of codes are encountered in computer systems. All should be planned well in advance of systems design, and most are included in the Standards Manual of the using organization. The types of codes are as follows:

1. Data codes—codes which represent values or characteristics of data.
2. Systems codes—codes which name files, programs, and data items.
3. User codes—codes which are used throughout the organization, usually to denote operating areas, products, departments, or levels.

Data codes: In many manual systems, and in all computer systems, codes are used to represent data for many reasons but primarily to reduce it to a manageable size for preparation and storage. If a college student is single, a freshman, a male, a resident of Hall dormitory, and studying to be a doctor, all this information could be stored as 11136 based upon a coding scheme where:

1 = single	1 = freshman	1 = male	1 = off campus	1 = Liberal arts
2 = married	2 = sophomore	2 = female	2 = athletic dorm	2 = Business
3 = divorced	3 = junior		3 = hall dorm	3 = Education
4 = widowed	4 = senior			4 = Agriculture
				5 = Engineering
				6 = Pre-med

This code requires minimum preparation time and can be stored on tape or disk in only five character locations.

Data codes, with a few exceptions, are used mostly by computer personnel, and the manager of this group should have the authority and responsibility for preparing the coding schemes in writing and distributing them to all personnel concerned. The structuring of data codes helps to show how important it is for systems analysts and designers to be concerned with an overall systems view rather than just the current systems design project. If, for example, another code used m = male and f = female instead of 1 = male and 2 = female, then the two codes would be incompatible and would not fit into the total system of an organization. Proliferation of minor code differences such as these will cause major problems when a system begins to operate.

Systems codes: The naming of programs, files, data items, and sometimes documents is the job of the systems analyst. As with data codes, there should be standards in this area for the systems team to follow, and a logical approach should be taken when creating these names. A typical structure might be as follows:

System
  Subsystem
    Application
      Input/output/program
        Numerical sequence

An example of a name that uses this structure is INVCUP01, where:

INV	stands for Inventory System
C	stands for Cost Accounting Subsystem of the Inventory System

U   stands for Update (other codes could be E = Edit, P = Print,
           etc.)
P   stands for Program (another code could be I = Input Records)
01   means that this is the first program in the Subsystem.

The use of these codes contributes to better control of the development of a
system and of its operation when implemented.

User codes: Any code used within an organization to denote some
characteristics of the organization or for any other reason may be considered a user
code. These codes are usually not determined by nor are they under the control of the
computer personnel; in fact, many of them may have been in use for years before a
computer was ever installed.

Examples of this type of codes are:

A = Executive   RXE = Assembly line   H = Home office
B = Supervisor   RXI = Inventory   E = East coast
C = Hourly worker   RBL = Utility and   W = West coast
                            steam plant

The first job of the systems team is to become familiar with all of these codes, and
if they are not in a Standards Manual or in writing, then the systems team should
prepare a list and distribute it to all persons concerned with the design project. Some of
the user codes will probably be quite logical and may be incorporated into the systems
design with little change, whereas others may be inefficient coding schemes for
computer use. Through an interchange of ideas with managers and users the systems
team may revise the codes, or they may be used as they are for some organization
reason. A good example of a user code would be the Library of Congress book
classification method.

Most considerations in the designing of codes should be covered in a Standards
Manual. However, if one is not available, the following are techniques and methods
which the designer must decide upon in almost all data systems.

Dates: Because of the difficulty in comparing dates when month, day, year
(1/6/77) is used, many computer systems use a Julian date. This is a system using 5
digits. The first 2 digits are the year and the last three digits are the day of the year. For
example, the sixth day of Janurary, 1977, would be 77006. December 25, 1977, would
be 77360. Although the Julian dates makes it easy to compare one date with another,
conversion routines must be available to print out the date in a readable form.
Whatever the method used, all date coding methods should be the same throughout the
entire system.

Time: It is suggested by most authorities in the systems area that a 24-hour
clock method be used for time. This makes it easy to distinguish A.M. from P.M. In the
24-hour clock method each hour is extended after 12 noon. 1 P.M. becomes 1300, 6
P.M., 1800, etc.

Industry terms: In situations where the computer system being designed is
to be used in a trade area or industry where special codes are in effect, it is usually better
to adopt the industry codes than to try to develop a new coding system. The systems
project manager should be responsible for distributing to the systems team any special
coding established within the industry for which the system is being designed.

Standard measurements: In data systems where standard measurements are
calculated, it is best to have a conversion routine that converts and stores all data in

decimal form. Inches, pounds, quarts, etc., are much easier to calculate and to maintain accurately if they are so converted.

Acronyms: Acronyms are words formed from the initial letters of a name or series of words. The government uses these words quite a bit and they are also found quite often in industry. USAF is the United States Air Force. UNICEF is the United Nation's Children's Education Fund. Smith and Appleton Corporation might be called SMIAPCO. When designing a system for the computer, it is better to use established acronyms for codes than to try to develop new ones.

Alphabetics: Although computer systems have the capability of printing or storing both upper and lower case alphabetics, a coding system should use upper case only. Also, certain alphabetic characters which are similar to numeric characters should be avoided in a coding scheme. The alphabetic O is often mistaken for the numeric 0 (zero). The hand printed S can be mistaken for a 5 (five). One (1) and I are often interchanged. Any alphabetic coding system developed for a computer should avoid I,L,O,Q,T,U,V, and S because they are often mistaken for each other or they are mistaken for numerics.

Numerics: Any numeric coding system with more than nine entities should use an expandable system. It is fine to code 1 = male and 2 = female; however, five types of liquid containers and six types of powder containers should not be coded 1 through 11. They should be coded 1 = liquid with 11, 12, 13, 14, and 15 for the five types, and 2 = powder with 21, 22, 23, 24, 25, and 26 for the six types. In other words, leave room for expansion of numeric digits.

Subdivisions: All coding schemes should consider whether a system needs to be subdivided and if so, to what extent. Subdividing coding schemes is similar to the subdivisions of a book or a theme.

Gaps: No coding scheme should use blank gaps. In other words, do not code 11 02; use 1102. Blank gaps may be put into display of codes but should not be used internally for storage, input, or comparisons.

Conciseness: In all coding systems, conciseness should be considered. Never use four digits where three digits will suffice.

SUMMARY

Many major decisions must be made by the systems designer about specifications of data and data management techniques. The systems analyst is concerned with how data is structured, how it is organized, how it should be accessed, how it should be processed, and how it may be standardized, secured, and controlled.

Proper use of most data is established by the Standard Data Management System which the manufacturer supplies. Within this system are many prewritten program routines called macros which will carry out organization, access, and file support of the data.

Besides specifying the methods of access, the systems designer must specify characteristics of each data item such as length, whether it will be blocked or not, what type of labels and keys will be used, and so forth.

There are a number of options open to the systems analyst and designer when he must select storage media and storage techniques. By carefully studying what is available to the particular computer system involved, he will find that these decisions become easier. Since most of the techniques of input, output, control, and procedures and logic are based upon how data is stored and accessed, most careful attention must be given to selection of file organization and access methods.

As a computer system becomes more flexible, Data Management Systems become more complex. Therefore, one of the basic jobs of the systems analyst and designer is to keep updated on the latest data management concepts and methods and to understand in detail how they affect the particular computer system with which he is dealing.

Questions

1. Name six objectives of data management.
2. What are the rules and conventions of data management used by the systems design team based upon? Where does the data management system come from?
3. Explain what the three parts of a data management system contribute to the system design effort.
4. Name the five levels of data and explain each one.
5. Discuss the advantages or disadvantages of creating a central data base for a system.
6. How can the need for immediate response determine the organization and access methods to be used in a system? How can the use of certain programming languages determine the organization and access methods to be used in the system?
7. Consider a simple file of employee records. Each record consists of a Social Security number, name, address, city, state, zip code, hourly rate of pay, sex and marital status. Using the 16 considerations presented in this chapter as a guide, discuss how you would lay out this file. Then answer the following questions:
   (a) How might the file be used in the organization?
   (b) What codes would you use for the data?
8. If the employee file just mentioned contains 400 records and is used to produce mailing labels for company literature, how would you organize it? How would you access it? How would you update it? How would you sort it?
9. If this same employee file contains 10,000 records and is used in a system where a manager may type the Social Security number into a terminal and obtain any employee records, how would you organize it? How would you access it? What would be the key?
10. If this same employee file contains 300,000 records of post office employees and each month about 4000 employee records must be updated, how would you update it? How would you organize it?
11. Explain the special considerations which may have to be given to user codes.

Exercises

1. Make a list of 15 characteristics of yourself, such as name, birth place, college attended, major area of study, military service, address, phone number, etc. Use a file record layout sheet to represent your record and the data items contained in it. On a separate sheet list your thoughts about the seven viewpoints discussed in this chapter in the section entitled Systems Design Considerations for Data Management. Assume that you will have 10,000 records similar to your own in your file. On another sheet of paper explain the coding scheme which you will use for the data items in such records.

2. Calculate the blocking factor you would use for your file given the following information:

    Memory locations available total 16,000.

    Operating systems requires 2,000 locations.

    Program requires 9000 locations.

    Input record is 65 characters.

    Output record is 110 characters.

    Double buffering will be used.

6

Output

Anything which exits from a system or is produced by a system is considered to be output. The output of a system is the result of processing the input using the functions of control and feedback. It may take many forms; common outputs of information systems are management reports, printed forms, displays, microfilm, and graphical drawings.

The production of output by the system is essentially the meeting of the objectives of the system. In other words, the objectives of the system are the outputs of the system. Therefore, output should meet all objectives defined by the managers and users in the systems study and should be the solution to the problem which is defined in the systems analysis phase. Determining what output is required essentially means the analysis of the objectives of the system, and the criteria for determining the quality of output is how well the output meets the objectives.

Output is always analyzed and specified before input because only by understanding exactly what is to be produced by the system can the analyst determine what will be needed as input.

TYPES OF OUTPUT

Output of an information system can be placed into three categories: (1) output as a basis for decision making, (2) output as a requirement to meet a functional objective, and (3) output as a requirement by law. These three categories have various degrees of importance; however, one of the basic needs of management is for information which aids decision making. No outputs of an information system have real value unless they are acted upon; that is, creating input, processing input, the control and feedback methods used, and the output created by the system as a basis for decision making have no value unless a decision is made. Only the action caused by the decision has value. Therefore, the value of output depends upon what action takes place that is based upon the output. It is common sense not to produce any output which is not used. However, much of this type of output is produced daily by computer systems now in use. Failure to analyze output requirements in light of problem solutions and meeting objectives is the main reason for this creation of unused output.

Most decision making based upon output of a computer system is made from reports produced by the system or by answers to inquiries into the system. These types of outputs will be considered in detail later in this chapter.

The second need for output is to meet requirements for some functional objective within an organization. Examples of these types of requirements are customer billing, bills of lading for shipping, purchase orders for obtaining raw materials, accouting reports, and many other types of outputs used to meet the objectives of the organization. Invoices which are sent to customers are probably never looked at by management and are not a basis for decision making. However, they remain a required output to be produced by the system.

The third output requirement is a requirement by law. All organizations are required to produce a certain amount of reports and forms for federal, state, and local governments. Examples of this type of output are W-2 forms, income tax forms, and special reports requested by government agencies.

DESIGN OF OUTPUT

The design of output essentially involves a creation of whatever is required to meet the objectives of the system. Consider the output report produced in Fig. 6-1. By analyzing this report, questions can be asked that the systems designer must answer. These questions are: (1) What data is needed to produce this report? (2) Is the report produced from some type of transaction? (3) Is it extracted from a master file? (4) Is the report cyclic in nature, that is, does it occur daily, weekly, monthly, or whatever? (5) Does the report meet the needs of the user? By discussions with the users of these outputs, information is obtained by the systems designer as to what he will need to consider when designing output. Some general considerations which should be thought of during output design are the elimination of unneeded information and the appearance of the output.

Another consideration of the systems designer is the creation of outputs that are desired. What constitutes a desirable output is debatable to some extent; however, if the output is to serve the need of the user the output should be what is desired by the user and not necessarily what the systems designer believes to be the most efficient for the system. Disagreements in these areas should be discussed, of course, and if no solution is found, the project manager or higher management must make a final decision. Misunderstanding about the capabilities of a computer system is sometimes a problem when designing output for users. Nontechnical users sometimes do not understand what computers can and cannot do. Also, it is sometimes difficult to describe to a nontechnical person exactly what the economics of computer use are and how they must be considered when producing certain types of information. The systems team should make every effort to acquaint all managers and users with the capabilities of the particular computer system which will be used by the proposed system.

Through this interaction of systems designers, users, and managers, information which is decided not to be needed should be eliminated.

TYPES OF REPORTS

Information systems produce many types of reports. They may be produced only one time or they may be cyclic in nature; that is, they may be produced daily, weekly, monthly, quarterly, or yearly. One consideration in the design of output reports is scheduling production so that it is not all due at one time, for example, at the end of the month. Not doing so would amount to inefficient use of the computer since it

INVENTORY STATUS REPORT

ITEM NO.	DESCRIPTION	QUANTITY ON HAND	QUANTITY ON ORDER	TRANSACTION QUANTITY	QUANTITY B/O	AVERAGE UNIT COST	EXTENDED COST	LAST RECEIPT	LAST ISSUE	MIN. BAL.	MAX. BAL.
411116	B500 TWINLITE SOCKET BLUE	458	500			.35	160.30			800	1600
	ADJUSTMENT			42		.35	14.70				
	RECEIPT			500		.37	185.00				
	ISSUE			50-		.36	18.00-				
		950*				.36	342.00	2/11/--	2/14/--		
411122	B506 SOCKET ADAPTER BROWN	325				.19	61.75			300	800
	ISSUE			20-		.19	3.80-				
	ISSUE			38-		.19	7.22-				
	ISSUE			10-		.19	1.90-				
		257*				.19	48.83	12/19/--	2/11/--	UNDER	
411173	C151C SILENT SWITCH IVORY	50	150			1.16	58.00			100	200
	RECEIPT	200*	150	150		1.20	180.00				
						1.19	238.00	2/10/--	2/03/--		
411254	A210 PULL CORD GOLD	62	75			2.25	139.50			80	165
	ISSUE			16		2.25	36.00				
	ISSUE	16*	75	30		2.25	67.50				
						2.25	36.00	11/17/--	2/10/--		

FINAL TOTALS
BEG. INV 48295.26
CHANGE 700.08
NEW VALUE 48995.34

Fig. 6-1 Output report (Courtesy, IBM Corp.)

would be very busy at one particular time of the month and very little used at other times. The reporting times should be spread out not only to get more extensive and more efficient use of the computer but also to avoid delays in producing output.

Types of reports that might be produced by a system can be categorized as follows:

1. Detail reports
2. Exception reports
3. Periodic reports
4. Management reports
5. Inquiry reports
6. Information listing reports

Detail reports: A detail report produces one line of output or one record of output for every record of input. Stated in simplier terms, a detail report lists all information available from that particular file. An example of the detail report would be a listing of every item in inventory, the amount on hand, the cost of the items, etc. The detail report may be summarized and broken up into various groups (Fig. 6-2). However, they are usually just long reports used for reference purposes.

Exception reports: The exception report produces information about unusual conditions. Examples of exception reports are a list of all customers who are overdue in paying their accounts, a list of all customers who have not purchased from the organization in the last six months, listings of materials which have been purchased but not delivered, listings of personnel who work excessive overtime hours, etc. Exception reports are a vital portion of the reports needed to make a management information system. The exception report is a report which management can immediately use as a basis for decision making.

Periodic reports: Reports which are periodic in nature can take many forms. Basically, a periodic report keeps personnel informed, interested, and involved. Such reports indicate the progress of an organization, for example, comparing current data with past data. Cost accounting reports sometimes fall into this category since they are not actual financial reports but rather reports on variations of expenses and income within an organization.

Management reports: Those reports directed toward management are known as management reports. Such reports inform a manager about his own area or about the general condition of the organization as a whole. Examples are reports to foremen on the assembly line about the number of hours worked by personnel and the use of raw materials and machinery, cost accounting reports on a profit center, and any other type of report which is directed toward a specific manager and his specific area. Usually, these reports take the form of summaries similar to the report in Fig. 6-3. This type of reporting is sometimes referred to as data reduction in the sense that the data is reduced to its most concise and informative form.

Inquiry reports: An inquiry about data which is stored within a computer system results in an inquiry report. This type of report is produced by the writing of a program that was not planned and could not be known during systems design. Since these reports are sometimes used only once, they are expensive to produce. RPG (Report Programming Generator language), FMS (File Management Systems), and other information retrieval systems can be used to produce these reports. By means of them, specialized information which would not normally show up on any other type of report can be requested and produced by a special programming effort.

LAURENTIAN INDUSTRIES, INC.

COMPARATIVE ANALYSIS OF SALES BY ITEM

PERIOD ENDING 10/31/--   PAGE

ITEM NO.	DESCRIPTION	CURR. PERIOD QUAN. THIS YR LAST YR		PCT CHG	YTD QUANTITY THIS YR LAST YR		PCT CHG
624634	D20068 OVERHAUL GASKET	10	14	29-	90	98	8-
624832	17D0011 BELT DYNAMIC FAN	190	150	27	1,820	1,905	4-
624901	DMK6448 HUB ASSEMBLY J2	1-	5	120-	18	18	0

LAURENTIAN INDUSTRIES, INC.

SALES BY ITEM CLASS

MONTH ENDING 03/31/--

ITEM CLASS	CLASS DESCRIPTION	SOLD THIS MONTH	GROSS PROFIT	PROFIT PERCENT	SOLD THIS YEAR	GROSS PROFIT	PROFIT PERCENT
1	ABRASIVES	2,720.19	271.36	10	9,900.17	907.60	9
2	ACIDS AND CHEMICALS	1,216.27	170.27	14	3,139.68	408.07	13
3	BRASS	6,220.83	435.45	7	16,341.47	1,143.87	7

LAURENTIAN INDUSTRIES, INC.

COMPARATIVE SALES ANALYSIS

BY ITEM CLASS FOR EACH CUSTOMER

MONTH ENDING 05/31/--   PAGE

CUST NO	ITEM CLASS	CUSTOMER/ITEM CLASS NAME	MONTHLY SALES THIS YEAR LAST YEAR		PRCNT CHG	YEAR TO DATE SALES THIS YEAR LAST YEAR		PRCNT CHG
3310		TARDELL HARDWARE						
	11	BUILDER HARDWARE	103.19	91.31	13	515.92	729.43	29-
	12	ELECTRICAL SUPPLIES	87.58	85.02	2	435.57	375.29	16
	13	GIFTS AND SUNDRIES	63.01	.00		315.09	490.36	35-
	14	HOUSEWARES	198.05	150.23	32	990.32	1,123.19	12-

LAURENTIAN INDUSTRIES, INC.

COMPARATIVE SALES ANALYSIS BY CUSTOMER

FOR EACH SALESMAN

PERIOD ENDING 07/31/--   PAGE

SLMN NO.	CUST. NO.	SALESMAN/CUSTOMER NAME	THIS PERIOD THIS YEAR	THIS PERIOD LAST YEAR	YEAR-TO-DATE THIS YEAR	YEAR-TO-DATE LAST YEAR
10		A R WESTON				
	1426	HYDRO CYCLES INC	3,210.26	4,312.06	10,010.28	9,000.92
	2632	RUPP AQUA CYCLES	7,800.02	2,301.98	20,322.60	11,020.16
	3217	SEA PORT WEST CO	90.00CR	421.06	900.00	593.10
		SALESMAN TOTALS	10,920.28	7,035.10	31,732.88	20,614.18
12		H T BRAVEMAN				
	0301	BOLLINGER ASSOCIATES	100.96	0.00	100.96	0.00

Fig. 6-2   Output report (Courtesy, IBM Corp.)

PLANNED EXPENSE ANALYSIS

A FUNDS SUPPLYING DEPT. 599

	CURRENT MONTH					YEAR - TO - DATE				
	ACTUAL EXPENSE	INCREASE OR DECREASE — FROM LAST MONTH		LAST YEAR		ACTUAL EXPENSE	INCREASE OR DECREASE -- FROM PLAN		LAST YEAR	
701 SALARIES - OFFICER	8,104.13	270	3.4%	395-	4.6%-	15,937.43	1,414-	8.1%-	1,062-	6.2%-
704 SALARIES - EMPLOYEES - REGULAR	1,280.00	132-	9.3%-	90	7.5%	2,692.50	149	5.8%	676	33.5%
705 SALARIES - EMPLOYEES - PART-TIME							36-		70-	
706 SALARIES - EMPLOYEES - OVERTIME	5.40	5		1	27.6%	5.40	5		1	27.6%
710 BONUS PROVISION	972.00			19-	1.9%-	1,944.00			39-	1.9%-
711 CAFETERIA OPERATION - ALLOCATED #	72.00			30	71.4%	144.00			60	71.4%
712 CONTRIBUTIONS UNDER PENSION PLAN	1,337.28			529-	28.3%-	2,674.56	1		1,058-	28.3%-
713 GROUP LIFE INSURANCE	40.35			52-	56.3%-	80.48	2-	2.6%-	103-	56.1%-
714 HOSPITALIZATION INSURANCE	55.48	3	6.2%	2-	4.3%-	107.72			7-	6.7%-
715 SOCIAL SECURITY TAX	340.39	5	1.5%	11-	3.1%-	675.58	266	65.1%	16-	2.3%-
717 UNEMPLYMENT INSURANCE TAX		166-				166.89			98-	37.1%-
718 OTHER DIRECT EMPLOYEE EXPENSES				25-			51-		25-	
719 EMPLOYEE BENEFITS DEPARTMENT - ALL #	20.94	11-	36.3%-	12		53.84	14	35.2%	38	
720 PAYROLL DEPARTMENT - ALLOCATED #	40.07	2	7.7%	8	25.0%	77.26	20	36.4%	10	15.4%
721 PERSONNEL DIVISION - ALLOCATED #	98.28	5-	5.0%-	45	87.0%	201.79	8	4.3%	98	95.1%
730 OTHER PROFESSIONAL FEES	56.33			56		112.66			112	
740 MACHINE RENTAL		50-				50.80	50		50	
742 MAINTENANCE & REPAIR-FUR. FIX & EQ	266.20	185		266		346.96	309		271	
745 PRINTING - ALLOCATED	8.64	4-	31.8%-	7		21.31	12		19	
747 PRINTING & OFFICE SUPPLIES	75.79	58-	43.6%-	44-	37.0%-	210.40	88-	29.6%-	154-	42.4%-
748 SUPPLY DEPT. - ALLOCATION #	6.64	6-	49.5%-	1	23.4%	19.80	6-	25.2%-	1	9.1%
753 BUILDING RENTALS-OWN 4,230 SQ. F	2,115.00			1,474		4,230.00	1		2,948	
759 DEPRECIATION-FURN. & EQUIP.	57.31			23	68.8%	114.62			46	68.8%
772 MAIL DEPT. & POSTAGE ALLOCATION	41.41	2-	5.6%-	9	27.8%	85.31	11-	11.5%-	6	8.4%
773 POSTAGE AND EXPRESS		1-		12		1.25	1		25	
775 TELEPHONE - TOLL CALLS	259.75	40	18.2%	101	64.2%	479.48	129	37.1%	80	20.2%
777 TELEPHONE - PBX DEPT. - 14 PHONES	280.00			60	27.2%	560.00			120	27.2%
780 TELEGRAPH DEPT. - ALLOCA 4 WIRES	2.88	1	65.5%			4.62	4-	47.6%-	2	
784 ADVERTISING - BUDGETED	2,436.60	2,436		309-	11.2%-	2,436.60	3,083-	55.8%-	2,773-	53.2%-
787 AUTOMOBILE EXPENSE		88-		32-		88.65	53		36	68.4%
792 DINING ROOM COSTS - ALLO 151 MEALS	604.00	240	65.9%	284	88.7%	968.00	234	32.0%	192	24.7%
794 MEMBERSHIP DUES	76.00	572-	88.2%-	5	7.0%	724.00	167	30.0%	469	
796 ENTERTAINMENT	436.14	255		238		617.01	293	91.0%	184	42.6%
797 TRAVEL	2,051.02	345	20.2%	549-	21.1%-	3,756.56	150	4.1%	639	20.5%
807 GARAGE DEPT - ALLOCATION #	1,196.44	19-	1.6%-	872		2,412.46	38	1.6%	1,970	
812 PUBLICATIONS & SERVICES				228-			92-		253-	
814 SURETY BOND PREMIUM	21.82			9-	29.9%-	43.64			18-	29.9%-
820 MISCELLANEOUS	165.00	155		112		175.00	110		98	
844 CREDIT DIVISION #	5,832.42	1,263-	17.8%-	2,118	57.0%	12,927.93	1,250	10.7%	3,926	43.6%
COLUMN SUB-TOTALS	28,355.71	1,562	5.8%	3,623	14.6%	55,148.51	1,518-	2.6%-	6,406	13.1%
880 ADMINISTRATIVE DEPARTMENTS	3,007.69	890	42.0%	631	26.5%	5,125.12	1,666	48.1%	1,362	36.2%
882 TAXES ON CAPITAL	3,235.38	156	5.0%	161-	4.7%-	6,314.67	241-	3.6%-	127-	1.9%-
COLUMN TOTALS	34,598.78	2,609	8.1%	4,093	13.4%	66,588.30	92-	.1%-	7,641	12.9%
NON-CONTROLLABLE TOTALS--#	7,266.79	1,303-	15.2%-	3,088	73.9%	15,837.08	1,325	9.1%	6,105	62.7%
CONTROLLABLE TOTALS	21,088.92	2,866	15.7%	534	2.6%	39,311.43	2,843-	6.7%-	300	.7%

Fig. 6-3 Management report (Courtesy, *IBM Corp.*)

Inquiry reports may also be thought of as information requested in an immediate response system. For example, when an inquiry is made whether there is a room available at a certain hotel, the computer output constitutes an inquiry report.

Information listing: A printout of data based upon specialized criteria is called an information listing, such as the output of every checking account balance of customers of a bank. This listing could be kept at branch banks so that account balances for all persons who have checking accounts can be referred to. Information listings are usually stored in designated work areas for constant reference by personnel. They become an integral part of a system and are produced at periodic times so that the exact reference needed is always available.

Various types of output reports are illustrated in Figs. 6-4 through 6-9.

OUTPUT SPECIFICATIONS

All output from a system should be specified in writing, usually by the use of special forms such as record layouts, printer layouts, forms layouts, and specialized listings. All data items which are part of the file are shown in the record layout. Record layout sheets are circulated by the systems team so that reference to any file can be made at any time.

A printer layout sheet is shown in Fig. 6-10. Printer layouts, sometimes called print charts, are samples of how a printed report will look. These printer layouts are used for the interchange of ideas between the systems designer and the user of the report. They are usually prepared in pencil so that they can be easily changed and updated according to the desires of the user and the needs of the system.

Forms layouts are needed by the form manufacturer for producing a sample which can later be approved and produced in volume. The preparation of forms layout is discussed in more detail later. Specially designed forms of this sort are often needed.

Output to be produced usually depends upon the following considerations:

1. The reason for it
2. The data needed to produce it
3. The format
4. The volume
5. The frequency

Deciding the reason for output: Before any output is included within a report, there must be a basic reason for it to be in the report. Sometimes this is a common sense decision which needs very little discussion. In other cases, certain types of output must be justified by the systems designer, the users, and the managers. Every data item included as output should have a written statement for its inclusion.

Data needed to produce output: The data needed to produce a data item requires another written specification, which should be prepared by the systems designer. For every output data item to be produced there should be a statement confirming the input data item or items required to produce it. These written specifications are essential to the programmers when they prepare their programs.

Format: The format of a data item is stated in the written specifications. Many times the formats that are stated on the record layout and printer layout forms are adequate; however, sometimes a detailed statement of format will be needed. Format statements usu lly include such information as whether a data item will be alphabetic, numeric, al hanumeric, edited with dollar signs, commas, etc.

LAURENTIAN INDUSTRIES, INC.

AGED TRIAL BALANCE

PAGE 1

DATE 7/17/--

CUST. NO.	CUSTOMER NAME	TELEPHONE NUMBER	LAST PAY MO DA YR	CREDIT LIMIT	TOTAL OUTSTANDING	CURRENT AMOUNT	OVERDUE ACCOUNTS		
							30 DAYS	60 DAYS	90 & OVER
108	ALLEN & CO.	415-378-1089	2 16 --	$15,000	$ 7,296.35	$ 6,919.77	$ 376.58		
165	ANDERSON AUTO SUPPLY	408-286-6741	1 28 --	2,500	1,665.49	1,665.49			
178	ANDREWS AND SONS INC	408-262-2074	2 05 --	750	146.64			$ 146.64	$ 90.44
F189	ARGONAUT ENGINEERING	415-867-2506	12 27 --	2,000	3,125.41	2,111.30	611.54	312.13	51.00
247	BERKLEY PAPER CO	408-251-4189	2 21 --	6,300	5,289.00	1,185.50	2,652.45	1,400.05	762.19
252	BEST DISTRIBUTION CO	408-296-1667	10 06 --	1,000	765.44	3.25			
	FINAL TOTALS				$35241.33•	$21,085.31•	$5,601.57•	$3,831.82•	$998.63•

Fig. 6-4 Output report (Courtesy, IBM Corp.)

Note: Letter F indicates that customer 189 is overextended on his credit line.

STATE OF NEBRASKA - DEPARTMENT OF REVENUE
INDIVIDUAL INCOME TAX ESTIMATED ERROR LISTING
PAGE 20

SOC SEC	SPOUSE SS	STATE ID	BATCH	REP	TC	TSC	DATE	TY	PP	BEG	END	-EXEMPTIONS-	AMT REC	LINE 10C	PREV PAY REM EST
521289878	508608112	15744442	14101	583	1		70105	69	18		1 1	2	35756	62500 NO ESTIMATE	
52307339A	505281134	15770273	12433	310	1	202	70104	69	15		1 1	2		2500 NO ESTIMATE	
523225755	50628A812	18045138	23461	100	41	112	70203	69	30					4900 NO ESTIMATE	
523288621	523288621	15776050	17336	339	1		70105	69	21	1	6	8	17113	20000 ESTIMATE ERROR	15000
52346974A		15780872	23764	206	2		70217	69	32	1	3	5		34340 NO ESTIMATE	5000
524030052		15785300	17354	26	1		70105	69	20	1	3 1	6		157200 NO ESTIMATE	
524124280		1578721A	17233	468	1		70105	69	20						

STATE OF NEBRASKA - DEPARTMENT OF REVENUE
INCOME TAX REFUND CLAIM REGISTER
PAGE- 1

PRIME SOC-SEC NO.	STATE IDENT NO	SPOUSES SOC-SEC NO.	BATCH NUMB	RPT NO.	TRAN CODE	TAX YR	PROC PRO	CLAIM ID NUMBER	NAME	REFUND AMOUNT	STATUS CODE
030-38-5209	18044212	000-00-0000	24081	192	1	69	33	614418	EMBDEN/CLARA J	22.00	
254-56-0558	10108211	000-00-0000	23970	405	2	69	32	614419	MCCOMMON/BESSIE	12.00	
429-70-7814	10304770	000-00-0000	24045	54	2	69	33	614421	COOPER/MITCHELL	12.16	
432-32-6875	10317198	000-00-0000	23961	237	2	69	32	614423	OGLESBY/MELVIN M	14.00	
446-42-0885	16506766	000-00-0000	23991	198	2	69	32	614427	ALBRIGHT/KEITH A & SHARI M	64.40	
478-20-0728	10444947	000-00-0000	24038	92	1	69	33	614428	KANE/WALTER J & JOSEPHINE A	105.64	
485-56-0740	18043062	000-00-0000	23867		1	69	32	614429	NELSON/TERRY L	7.00	
530-30-8880	16971051	506-28-7605	23991	234	2	69	33	614430	BAUER/ROBERT L	52.00	
		000-00-0000	24096	19	1	69	32	614432	JOHNSON/LEROY E & MARIE J	32.31	
	10707603	000-00-0000	0186B	424	2	69		614433	WENDT/WILLIAM E & MILDRED G	5.01	
								614437	SAMUELSON/PAUL E & LORETA	31.35	
									SCHNEIDER/DALE M F ALICE H		

STATE OF NEBRASKA - DEPARTMENT OF REVENUE
INDIVIDUAL INCOME TAX BALANCE AND EDIT ERROR LISTING
PAGE 2

BAT REP	REC TRC LINE 1	DATE LINE 2	TY	PP	FC NS LINE 3	DATE ENT LINE 6A-C	SOR FS LINE 7	SOC SEC LINE 8	SPOUSE SS L-10A L-103	BEG L-103	END L-10C	EXEMPTIONS L-10D L-10B	TSC LINE 12	AMT REC LINE 13 LINE 14	LINE 15	EDIT ERROR
•••••																
20967	180		13	0•												BATCH NO 20967 IS IN ERROR
20967	180		13	0										2100-		
20967	180		13	0										2100-		UNMATCHED TRANSACTION

Fig. 6-5 Output reports produced by a system installed in the State of Nebraska

H.B. ZACHRY COMPANY
JOB NO. 561

CONSTRUCTION OVERHEAD COST STATUS
AS OF 31 MARCH 19
MATERIAL TESTING LAB
PAGE 24

COST CODES DISTR CODE	EXPENSE CODE	DESCRIPTION	TOTAL ESTIMATE	JOB STATUS TO-DATE ESTIMATE	ACTUAL	VARIANCE	UNUSED ESTIMATE	PERCENT OF ACT. TO EST. COST	JOB COST TRENDS PERCENT UNUSED ESTIMATE	PERCENT UNUSED TIME
10	01	GENERAL EXPENSES SALARIES AND WAGES	10000	2500	2550	50-	7450	102%	75%	75%
	04	PAYROLL TAXES	600	150	153	3-	447	102%	75%	
	14	FUEL, LUBE AND SERVICING	850	213	150	63	700	70%	82%	
	26	RENTALS-INTER-CO AIRCRAFT	200	50		50	200		100%	
	27	RENTALS-INTER-COMPANY EQUIPMENT	6000	1500	741	759	5259	49%	88%	
	51	SUPPLIES	100	100		25	75		100%	
	69	SERVICES NOT SUB-CONTRACTED	500	125	438	313-	62	350%	12%	
	73	TRAVEL AND SUBSISTENCE	1500	375	403	28-	1097	107%	73%	
	74	UTILITIES	100	25	1	24	99	4%	99%	
		TOTAL MATERIAL TESTING LAB	19850	4963	4436	526	15414	89%	78%	
561	TOTAL		19850	4963	4436	526	15414	89%	78%	

H.B. ZACHRY COMPANY

MONTHLY OPERATIONS SUMMARY-OVERHEAD COST
PERIOD ENDING 31 MARCH 19

COST CODES ACCOUNT	JOB	DESCRIPTION	TOTAL ESTIMATE	JOB STATUS TO-DATE ESTIMATE	ACTUAL	VARIANCE	UNUSED ESTIMATE	PERCENT OF ACT. TO EST. COST	JOB COST TRENDS PERCENT UNUSED ESTIMATE	PERCENT UNUSED TIME
700		ADMINISTRATIVE AND GENERAL								75%
	720	LEGAL	115710	28928	29160	232-	86550	101%	75%
	730	ACCOUNTING DEPARTMENT	150688	37672	52755	15083-	97932	140%	65%
	740	ELECTRONIC DATA PROCESSING	157776	39944	39606	161-	118170	100%	75%
	742	SYSTEMS DEVELOPMENT	80000	20000	25252	5251-	54748	126%	68%
	750	PUBLIC RELATIONS	67289	16822	16988	165-	50301	101%	75%
	760	EMPLOYEE RELATIONS	72635	18159	34147	3708	50488	93%	77%
	780	GENERAL EXPENSES CONSTRUCTION PLANT PROPERTY	72900	18225	9764	8460	63135	54%	87%

H.B. ZACHRY COMPANY

MONTHLY OPERATIONS REPORT
HEAVY CONSTRUCTION
PERIOD ENDING 31 OCT 19

JOB NUMBER	DESCRIPTION	TOTAL ESTIMATE	PCT. EST. COST TO TOTAL EST.	JOB STATUS TO-DATE ESTIMATE	ACTUAL	VARIANCE	JOB STATUS THIS MO. ESTIMATE	ACTUAL	VARIANCE	% ACT TO EST COST TO-DATE	THIS MONTH
410		HEAVY CONSTRUCTION										
1023		COUNTY HIGHWAY										
	LABOR	747340	58	434667	441225	6358-	97130	94368	2762	101	97
	EQUIP.	530494	55	293458	290102	3356	46254	48207	1953-	98	104
	MATL.	164586	53	102685	912485	6103-	20663	22478	1815-	93	108
	SUPP.	252201	53	100285	103269	10984-	20663	22478	1815-	111	
	SUB-CON	207154	39	97447	103487	6040-	18995	16233	2762	106	89
	TOTAL	3554646	53	1887957	1848668	39299	291507	292890	1383-	98	100
1638		AIRPORT RUNWAY										

Fig. 6-6 Output reports

MERCHANDISE INVENTORY EXCEPTIONS   PAGE 4

LEGEND OF ERROR CONDITION CODES
CODE -1 ABSTRACT QUANTITY VS. INVENTORY QUANTITY
    -2 ABSTRACT COST VS. INVENTORY COST
    -3 INVENTORY EXTENSION ERROR
    -4 INVENTORY COST VS. MASTER COST FILE
    -5 NC MATCHING COST MASTER RECORD
    -6 ITEMS CARRIED AT NO COST

CODE	COMM CODE	BRANCH	ABSTRACT	MSTR COST	INVENTORY QNTY	INV COST	EXT COST	DIFFR
4	1640458	12		3.650		3.230		.420 CR
4	1640467	8		.000		2.230		2.230
4	1640490	15		5.820		5.370		.450 CR
6	1640606	3		.000		.000		.000
6	1640673	9		.000		.000		.000
6	1640949	7		.000		.000		.000
5	1642024	18	NO U.O.M.		362.00	.000	.00	
5	1642025	5	NO U.O.M.		121.00	.000	.00	
5	1642026	18	NO U.O.M.		500.00	.000	.00	
5	1642027	11	NO U.O.M.		571.00	.000	.00	
6	1650410	3		.000		.000		.000
6	1650503	12		.000		.000		.000
6	1650505	2		.000		.000		.000
4	1650558	12		.000		1.880		1.880
4	1661110	14		5.150		4.930		.220 CR
6	1680550	1		.000		.000		.000
6	1680552	13		.000		.000		.000
6	1680553	4		.000		.000		.000
6	1680554	10		.000		.000		.000

Fig. 6-7   Output report

The Systems Analyst

```
FARM NO.  1234567890          PROFILE FARM ANALYSIS - DAIRY              YEAR  19--
```

CAPITAL EFFICIENCY FACTORS		YOUR FARM	LIKE FARMS	TOP FARMS
INVESTMENT/COW (14,000 LBS 3.5 MILK)				
TOTAL INVESTMENT	$	1,630	1,900	1,350
LAND AND BUILDINGS	$	650	650	390
MACHINERY AND EQUIPMENT	$	280	490	360
LIVESTOCK	$	500	570	440
INVESTMENT/TILLABLE ACRE				
LAND AND BUILDINGS	$	270	240	160
MACHINERY AND EQUIPMENT	$	120	180	150
INVESTMENT/MAN				
TOTAL INVESTMENT	$	64,000	52,000	51,000
MACHINERY AND EQUIPMENT	$	11,000	13,000	14,000
LIVESTOCK	$	20,000	16,000	17,000
CREDIT USE				
TOTAL DEBT/$1.00 OF INVESTMENT	$	.27	.32	.33
TOTAL DEBT/$1.00 OF INCOME	$	.50	1.08	.52
SHORT TERM DEBT/$1.00 OF INVESTMENT	$	.17	.14	.18
SHORT TERM DEBT/$1.00 OF INCOME	$	.32	.47	.29
LONG TERM DEBT AS % OF TOTAL DEBT	%	.63	.44	.55

```
FARM NO.  1234567890          PROFILE FARM ANALYSIS - DAIRY              YEAR  19--
```

SIZE FACTORS		YOUR FARM	LIKE FARMS	TOP FARMS
LIVESTOCK				
COWS	#	100	82	107
YOUNGSTOCK	#	69	51	72
RATIO OF YOUNGSTOCK TO COWS	%	69	62	67
LAND				
OWNED	ACRES	314	300	398
RENTED	ACRES	0	15	11
TILLABLE	ACRES	240	206	258
CROPS	ACRES	240	187	236
HAY	ACRES	121	77	93
HAY CROP SILAGE	ACRES	0	16	25
CORN SILAGE	ACRES	40	57	82
CORN GRAIN	ACRES	55	17	19
OATS	ACRES	0	12	5
OTHER GRAIN	ACRES	24	8	12
TILLABLE ACRES PER COW	ACRES	2.4	2.5	2.4
FORAGE CROP ACRES PER COW	ACRES	1.6	1.8	1.9
GRAIN CROP ACRES PER COW	ACRES	.8	.5	.3
CAPITAL				
LAND AND BUILDINGS	$	66,000	48,500	42,500
FEED AND SUPPLIES	$	20,500	13,800	17,800

```
FARM NO.  1234567890          PROFILE FARM ANALYSIS - DAIRY              YEAR  19--
```

PRODUCTION EFFICIENCY FACTORS		YOUR FARM	YOUR FARM	LIKE FARMS	TOP FARMS
MILK					
TOTAL SOLD	LBS		1,372,000	1,016,000	1,521,000
AVERAGE PER COW	LBS		13,700	12,400	14,200
BUTTERFAT TEST	%		3.75	3.70	3.60
BUTTERFAT SOLD PER COW	LBS		514	458	510
TOTAL INCOME FROM MILK	$		79,900	56,000	85,500
MILK INCOME PER COW	$		799	683	800
FEED					
PURCHASED FEED COST/COW	$		143	150	172
HOMEGROWN GRAIN VALUE/COW	$		76	34	26
ROUGHAGE VALUE/COW	$		135	136	127
TOTAL FEED VALUE/COW	$		354	320	325
TOTAL HAY EQUIVALENT/COW	TONS		7.8	6.4	7.0
FEED TO MILK CONVERSION INDEX			123	127	108
PURCHASED FEED AS % OF MILK INCOME	%		18	22	22
TOTAL FEED AS % OF MILK INCOME	%		44	47	41
LABOR					
COWS/MAN	#		39	30	38
MILK SOLD/MAN	LBS		518,000	375,000	525,000
GROSS INCOME/MAN	$		33,600	22,500	32,100

```
FARM NO.  1234567890          PROFILE FARM ANALYSIS - DAIRY              YEAR  19--
```

COST FACTORS		YOUR FARM	LIKE FARMS	TOP FARMS
MACHINERY				
REPAIRS	$	3,878	2,745	2,532
GAS AND OIL	$	1,808	1,346	1,420
UTILITIES	$	866	712	942
DEPRECIATION	$	2,840	3,640	3,930
INTEREST ON INVESTMENT	$	1,989	2,550	2,750
HIRED CUSTOM WORK	$	265	285	327
LESS CUSTOM WORK DONE	$	0	112	43
TOTAL	$	11,646	11,166	11,858
PER CROP ACRE	$	49	60	50

Fig. 6-8   Output report

INVENTORY
SUGGESTED ORDER LIST

KEY	CAT. NO.	DESCRIPTION	MAX. QUANT	ISSUE COST	AMOUNT ON HAND	AMOUNT ON ORDER	MIN. QTY.	SUG. ORDER	INVENTORY COST	LAST P.O. COST
7280	S METER	CLAMP 12 IN BELL JOINT 60S	3	27.5500EA	2		1		55.10	
7350		CLAMP 12X20 IN S&B FOR C I PIPE	4	86.2500EA	2		1		172.50	
7420		CLAMP 12X7 1/2 IN S&B FOR C I PIPE	4	38.0000EA	1		1	3	38.00	
7490	S METER	CLAMP 14 IN BELL JOINT 60	3	48.3500EA	2		1		96.70	
7560	S METER	CLAMP 14 IN BELL JOINT 60S	3	63.0000EA	2		1		126.00	
7630		CLAMP 14X20 IN S&B FOR C I PIPE	3	117.0500EA	1		1	2	117.05	
7700		CLAMP 14X7 1/2 IN S&B FOR C I PIPE	2	43.0000EA	2		1		86.00	
7770		CLAMP 16X20 IN S&B FOR C I PIPE	2	133.5800EA	1		1	1	133.58	
7840	S METER	CLAMP 10 IN BELL JOINT 60	3	21.2800EA	2		1		42.56	
7910	S METER	CLAMP 10 IN BELL JOINT 60S	3	24.0000EA	2		1		48.00	
7980		CLAMP 10 X 7 1/2 IN S&B FOR C I PIPE	2	33.0800EA	2		1		66.16	
8050	GRINNELL	CLAMP 2 1/2 X 3/4 IN DOUB STRAP	10	4.2100EA	6		2		25.26	
8120	GRINNELL	CLAMP 2 1/2 X 3/4 IN SING STRAP	10	2.8000EA	44		2		123.20	
8190		CLAMP 2 X 3/4 IN DOUB STRAP	20	3.7900EA	13		5		49.27	
8330		CLAMP 3 X 3/4 IN DOUB STRAP	10	4.3500EA		6	2		.00	
8400	GRINNELL	CLAMP 3 X 3/4 IN SING STRAP	10	3.5000EA			2	10	.00	
8470	S METER	CLAMP 4 IN BELL JOINT 60	4	10.3000EA	3		2		30.90	
8540	S METER	CLAMP 4 IN BELL JOINT 60S	6	13.0100EA	4		2		52.04	
8610	DAVIS	CLAMP 4 X 15 IN S&B FOR C I PIPE	2	31.0000EA	2		1		62.00	
8680		CLAMP 4 X 20 IN S&B FOR C I PIPE	1	42.3500EA			0	1	.00	
8750		CLAMP 6 IN BELL JOINT 60	20	13.8200EA	11		6		152.02	
8820	S METER	CLAMP 6 IN BELL JOINT 60S	18	16.8100EA	8		6		134.48	
9100	GROOVER	CLAMP 6 X 3/4 IN DOUB STRAP		5000EA	5		2		32.50	

Fig. 6-9　Output report

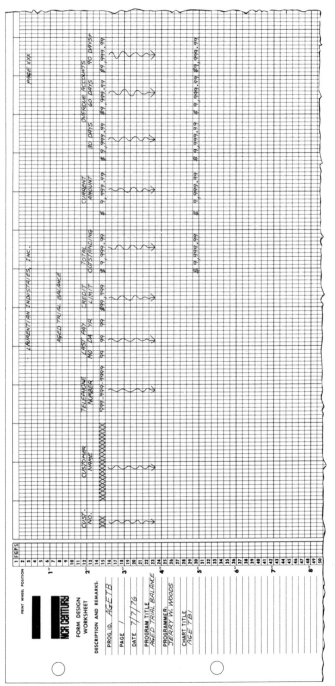

Fig. 6-10 Printer layout sheet

Volume: The volume of the output should be stated in approximate figures. Such figures are helpful in determining the length of computer runs, the number of forms that should be ordered, how much paper supply should be in stock, etc. Sometimes it is very easy to predict the volume. For example, if there are one-hundred workers in a company, then there will be one-hundred payroll checks. In other cases, it can be very hard to predict how much volume will be produced. However, some approximation should be stated if at all possible.

Frequency: The periodic frequency of the output items should be stated in writing. Basically, frequency of use determines the cost per unit of producing output. Obviously, it is much cheaper per unit to produce large volumes of output than it is to produce small volumes.

TYPES OF OUTPUT CODES

A computer system produces two basic types of output codes. First is the output which is to be read by human beings, usually called the report output or display output. The second type (called file output) is that output which is produced for computer storage. Such output will be used later as input.

Some of the basic differences between these types of output should be noted by the systems designer. Report output, for example, is produced to be as readable as possible. Therefore, there should be editing, leading zeros should be suppressed, the data should be of easily understood alphabetic or numeric form, and spaces should be used to separate columns of figures.

Output which is to be stored and used later by the system or some other system does not necessarily have to be produced in alphabetic or decimal form. It may be in binary form or it may be stored in a code system not readable by humans. There is usually no display editing of this type of output, leading zeros are not suppressed, and spaces are not used. In other words, readability is not a consideration of file storage output.

OUTPUT TECHNIQUES

The following is a list of output techniques which have been developed over several years of use within the computer industry. All of these techniques should be considered by the systems analyst as possible methods for assuring better readability, more accuracy, and optimum use of output.

1. Control totals
2. Sequence of output
3. Counting record numbers
4. Cross-foot checks
5. Editing techniques
6. Labeling
7. Calculation of subtotals
8. Limit checks
9. Use of proper headings
10. Conversion of codes to readable words
11. Historical checks
12. Use of page numbering

13. Spacing horizontally
14. Spacing vertically
15. Visual verification of samples
16. Use of output as input
17. Use of turn-around documents
18. Composition
19. Flags
20. Documentation

Some of these techniques have also been applied to input procedures, and there is some repetition of them in the next chapter. Moreover, the control of data and the control of output are considered in the chapter on "Control." However, these particular techniques are considered to be especially applicable to data in the output phase of computer runs.

Control totals: As processing takes place and output is produced, control totals can be derived. These control totals are usually printed out, perhaps at the end of the output report or upon the console typewriter of the computer operating panel. Control totals can be exact figures, or they may be approximations. For example, if the inventory value of an organization is approximately $1,000,000, any total figure for inventory which was much less or much greater than this figure would be subject to question. This particular figure may be considered as an approximate control total. An exact control total would be the 97 payroll checks needed for exactly 97 employees. Any figure other than 97 would be an error, and the output should be checked and corrected. Several kinds of figures produced at output time can be considered to be control totals. Sometimes a control does not have to be a total. If it is known that the last key number in a file is 12369, then the last key number printed by the computer should be 12369. Any type of number printed at the end of an output run which can be checked for exact or approximate accuracy can be considered a control over quality of output.

Control totals are not values which are determined at output time but rather are totals which have been carried through the system from creation of data to final output form. Control totals are also any total figures used in batching or editing of input data. The dollar total of all sales tickets might be a control over the printing of invoices; that is, the total amount of sales should equal the total amount printed on the invoices.

Sequence of output: The systems designer must pay careful attention to the sequence of output, whether report output or file output. As discussed before, report output usually is produced in sequence by some identification number, such as sales number, branch number, geographical location, or the like. The decision is made by agreement between the systems designer, the user, and the manager. Producing output in the proper sequence facilitates efficiency in using it and increases the readability of reports. See the chapter on "Procedures and Logic" for more information.

Counting record numbers: The process of counting the number of records within a file and producing this total at the end of an output run is a technique very similar to the control total. In many cases, either an approximate number or the exact number of the records which must be processed to produce output is known. Such numbers can be produced and checked at the end of output runs. For example, if it is known that there are approximately 50,000 items in an inventory list and the number of records counted is not approximately 50,000, it would seem clear that the listing is not complete or contains unneeded or inaccurate information. Record counts can also be obtained from trailer labels of the input files and compared with the number of records processed for output.

Cross-foot checks: Cross-foot checks are checks very similar to the accounting method of adding columns both horizontally and vertically. The totals derived must be equal. Cross-foot checks are also very similar to the zero balance technique used by accountants. In computer systems the particular numbers used do not have to be related numbers; therefore, any vertical total can be compared to any horizontal total to see if there is a zero balance or an exact equal figure for both totals. Checking of these totals during output is a reliable method for checking the accuracy of the data contained in the report.

Consider a report which produces, after much calculation, the values of A, B, and C. These three values can be added together to create a cross-foot total of D. The value of D would never have to be printed out and would usually be held in the computer's memory. Then, the total of all D's should equal the total of all A's, B's, and C's calculated and printed by the system. If not, then some type of error is indicated and can be found and corrected.

Editing techniques: Editing procedures contribute much to the readability of output reports. Examples of editing are the suppression of leading zeros, the use of dollar signs (the systems designer should be very conscientious in designing all dollar figures used on reports), insertion of commas, and the use of asterisks and other specialized symbols which may make the data item more readable and more presentable. Of course, editing procedures are determined by the needs of the users, and the edit is made only for readability. The editing of input data is an entirely different concept and is discussed in the chapter on "Input."

Use of labels: Most file media in modern computer systems contain labels at the beginning and end of the file. The use of these labels to control input is very common. Labels also are created during output. In some cases even printed reports should be labeled. The types of labels contained in file media such as magnetic tapes and disks are internal, magnetically recorded labels. Physical labels, those labels which appear on the outside of files such as gummed labels, are also important to control of output. Every type of data, whether file or printed media coming from the computer system, should be labeled immediately before or after it is created. Labeling helps in the distribution of output, as well as control of output to be used later in the system.

Calculation of subtotals: The calculation of subtotals can be used as an extra accuracy check of the output items. For example, subtotals of branch sales and home office sales of an organization can be compared with the total sales figure and provide a check very similar to cross-footing or the use of control totals. These subtotals can be printed out at the end of computer runs and compared against known major totals. In some cases, three or four different reports will produce three or four different subtotals. These subtotals can be compared with one known major total to check the accuracy of the reports.

Limit checks: The use of limit checks, a well-known input technique, can also be used on output. Any output value which is known not to exceed or be less than a particular limit can be checked by a parameter of limits. In other words, if a certain computation or number should always fall within three and six, it can be checked for less than three or greater than six. As exceptions to these limits are encountered during processing runs, they can be printed either on the console typewriter or upon the printer. The greatest use of limit checks is during creation of an output file where no printer output is produced. As the output file is created, a listing of exceptions should be produced which is called the "error register."

Use of proper headings: For output reports which must be read by working personnel or management, the use of proper headings is important. Abbreviations, codes, and other symbols should be avoided when possible. If more room is needed than is available on the output sheet, then several lines should be used to print the heading rather than trying to abbreviate it or squeeze it into a small space. Careful attention to placement of headings by the systems team and the programmers is important. There have been cases reported of values being produced in columns under headings which were not the correct headings for those values.

The use of proper headings contributes much to the readability of a report, makes communication between persons using reports much easier, and is always a good design and programming technique. A proper heading usually has a title at the beginning as shown in Fig. 6-7 and subheadings under the title to denote the columns produced. Headings should be repeated at the top of each printed page.

Converting codes to readable words: All codes used by the system should be converted to familiar English words when used on output records, if possible. Sometimes space does not allow for complete conversion of codes. However, it is much easier for the reader of a report if nonself-explanatory codes such as R for "reducing" and O for "oxidizing" are converted to the actual words. Codes such as 1, 2, and 3 standing for parcel post, air, and freight should be converted to abbreviations such as PP, AIR, and FRT, or to the actual words PARCEL POST, AIR, and FREIGHT on the output report. If codes cannot be converted, it is often possible to program the computer so that the codes on the report can be footnoted at the bottom of pages.

Historical checks: The use of historical checks upon output data is an old and valued practice. The computer can store historical data and check output against the historical data to see if the new output is reasonable. In other words, if a value is reasonable in view of the past performance of an organization, it will probably be reasonably accurate in the present. An example would be checks upon the total amount of payroll issued for the week versus the average amount issued in the last 20 weeks. Checks of this kind prevent large errors in a system such as those which produce checks for $1,000,000 when the amount was meant to be $100. Producing reports based upon historical checks is also a valuable contribution to management decision making. When a manager sees numbers which are out of line with past performance, he can make adjustments to correct the deficiencies in the areas involved. Historical checks are the basis of most cost accounting systems and can be used quite well in checking almost any type of current output value.

Use of page numbering: Usually, all pages produced by a system should be numbered. Preprinted forms are usually prenumbered by the manufacturer of the form. Use of page numbers assures quick reference to the report as well as a check upon the accuracy of the report. Many computer operators of some experience begin to sense when a report is of the correct size. For example, if the inventory listing usually runs 50 pages in length and the operator sees that it is now 70 pages, he can report this as a possible error. Another valuable technique is to have the operator enter the beginning number of any preprinted, prenumbered form and check this with the last number produced by the system. A computer system can add to the beginning number each time a prenumbered page is produced. The total produced by the computer system should be the same as the last prenumbered form.

Spacing horizontally: Correct use of horizontal spacing is important to the readability of forms. Figures should not be run together, and columns and numbers should be adequately spaced horizontally. Centering of pages and centering of headings adds to the appearance of the report. The analyst or designer usually makes decisions on horizontal spacing based upon the requests and comments of the users of the output.

Spacing vertically: Whether a report will be single-spaced, double-spaced, or tripled-spaced is a decision which the systems designer must specify. This decision is a trade-off between conservation of space and readability. The computer has the ability to either skip lines or space lines. Therefore, when large numbers of lines are to be placed vertically, they should be skipped rather than spaced. In other words, it is much more efficient for the computer to skip six lines at once than to space one line six times.

Visual verification of samples: For many output reports and for special reports of importance such as those which may be read by the company president, the board of directors, or some other high official within an organization, there should be specifications within the system for visual verification. Sometimes verification can be achieved by sampling techniques; other times personnel in the computer center or the user's section will visually verify the report. Verification does not necessarily mean that the report is recalculated. However, it does mean that certain figures are checked for accuracy and that the visual appearance of the report in general is also checked. When a computer produces 50, 100, or more pages of a report, each page should be checked visually if the report is considered to be a VIP type of report. Where customer relations and public relations are involved, visual verification can also be used to make sure that no output leaves the organization which is not accurate.

Use of output as input: A very important factor for the systems designer to consider is what output can be used later in his own system or in other systems as input. For example, invoices and other such documents returned to the organization may be used as input later on. Generally, every time the output produced can be used later as input there will be cost savings. Using the systems flowchart will help determine where output may be used again as input. Types of outputs which normally re-enter the system as inputs are punched cards, printed forms for optical scanning, and punched paper tape.

Use of turn-around documents: Turn-around documents are those documents produced by the system as output which can be later used in the system for input. Optical scanning techniques for optical character recognition (OCR) are the basis of many of these documents. There are other types of turn-around documents such as cards and paper tape. A card turn-around document is a card such as those sent with credit card billings, where the card is torn off and part of the card is sent back with the payment. This part of the card can be read into the system with minimum data preparation needed by the computer center staff. Turn-around documents, which are very simple in nature, should be used whenever possible.

Composition of information: Before a data item is used as output it should be checked for correct composition of its codes, characters, numbers, etc., to see whether numeric items are really numeric, that data items that should be present are not blank, and that codes contain the proper numbers of characters. These matters are usually taken care of by input editing; however, mistakes in writing programs can cause such errors to appear in output also. These checks can be made very easily. Any errors found should be printed on an error register or displayed to the computer operator.

Flags: An output flag gives some indication that a particular output item needs attention or is of an exceptional nature. A common flag consists of three asterisks in a row (***). These astericks are placed beside items that need attention. For example, on a report of customers' accounts, three astericks can be placed by the names of customers who are behind in payment to bring them to the attention of personnel. Flags constitute a very important output technique and should be considered by the systems designer for all printed output, not just exception reports or management decision-making reports.

Documentation: There should be written specifications for all output to be produced by a system. When personnel leave an organization, unless they have left written specifications behind as to how the system functions and what is produced, it will be very hard to make the system work effectively. Written specifications also provide for ease of communications between the systems designer and the users and managers of the system. Detailed methods of documentation are discussed in Chap. 10.

CONTROL OF OUTPUT

Control of output quality is accomplished by specifying proper output techniques. However, other important aspects of control which concern the systems designer include the following:

1. Written standards
2. Use of forms
3. Written distribution rules for each system
4. Checks and audits

Written standards: There should be written standards as to how output will be distributed, what security measures will be used, what personnel are responsible for distribution, methods of changing output requirements, approval of form changes, and a time schedule of when output is desired. Such standards are usually included in a book called the SOP, or Standard Operating Procedures. If no such book exists, the systems analyst will have to develop the standards.

Use of forms: The use of preprinted forms must be controlled so as to assure adequate supply as well as a distribution that is based upon the objectives of the company. The designer will probably need to develop written rules for the handling of special forms such as bank checks, which must be guarded closely to prevent fraudulent use and to protect against accidental loss, etc.

Written distribution rules for each system: As discussed in the chapter on "Documentation," there should be written distribution rules for each system implemented. For example, when an inventory listing is produced, a written document should state exactly how many copies are needed, the name of each person to whom the listing will go, and whether it is to be distributed by U.S. mail, office mail, hand carrying, or the like. Without written rules, chaos can result in distributing reports.

Checks and audits: There must be methods of checking the creation of output and methods of auditing its use. Use of log books and other methods discussed in the chapter on "Control" help to form an "audit trail" within an organization. The audit trail consists of those steps in data handling which can be traced from the creation of the data to its end use.

THE DATA REDUCTION SECTION

Within a computer center one section is usually made responsible for distributing output. This section is usually called the data reduction section, distribution section, control section, or some other such name. The devices of this section that should be considered by the systems designer are as follows:

1. Photocopy machines
2. Reproducing machines
3. Bursters and decollators
4. Mailing machines
5. Binding, printing, and other types of equipment

This section must maintain written specifications for the distribution of output. Its basic job is to receive printed reports from the computer center and distribute them to the proper destinations.

Almost every such section will make use of photocopy machines such as Xerox machines. If 100 copies of a report must be produced, it is not efficient for the computer to run multiple carbons over and over. The number of carbons that most computers can print legibly is limited to six to eight copies. Therefore, photocopying output is the method generally employed. Copies are then bound, stamped, folded in some fashion, and sent to their destination. Data reduction sections also sometimes use mimeograph and offset printing capabilities to reproduce reports in large quantities.

Carbon paper is removed from output forms by a device called a decollator. The pages of continuous forms are separated into single sheets by a device called a burster. Some equipment combines both functions. The use of these machines must be planned by the systems designer.

Mailing machines are used to distribute such data as payroll checks and reports which are going to other branches, warehouses, and factories.

The procedures which the data reduction section must follow are specified by the systems designer in his systems specifications. Flowchart symbols such as off-line operation or manual operation denote these operations very clearly. It can be seen that the design of output does not end when the output comes from the computer. The systems designer must also design the distribution of output and specify, in writing, the destination of each output and the proper procedures to ensure its getting there.

REAL-TIME OUTPUT

Another type of output to be considered by the systems designer is that which is immediately accessed from the system. This type of output is used in airline reservations systems and some management information systems. When a request is made for some type of information and the information is given immediately, this is termed real-time operation. Although there is much to consider in this area, real-time output basically relies on the output techniques discussed previously. One of the basic differences in real-time output is that there is usually interaction between the personnel requesting the output and the computer system. Therefore, the input which produces the real-time output must be specific. See Chap. 12 for a detailed discussion.

THE USER OF OUTPUT

Careful consideration must be given to the amount of computer output which leaves a system for distribution to system users. Oversaturation of users by output should be avoided, if possible. Computer systems have the ability to print up to 3000 to 4000 lines per minute. If they ran all day at this speed, more output might be produced than could be read by all persons in the organization. Oversaturation of output is sometimes called the wheelbarrow approach, alluding to the need for a wheelbarrow to deliver all the output which might be addressed to a particular person. The systems designer should be very careful that this does not happen. To avoid it, the use of summarization and exception reporting is recommended. As discussed above, the user should receive only what he needs and desires and very little else. Careful study of reports during the systems analysis phase can help eliminate the wheelbarrow approach.

FEEDBACK

The principle of feedback is very important in output design. Feedback is essentially the monitoring or measurement of the quality of output as a means of adjusting the performance of a system. Types of feedback include errors that lead to customer complaints, inaccurate payroll checks, and other conditions where comments, suggestions, or other means of communication are fed back to the persons managing and operating the system. The systems designer must allow for feedback channels and provide methods to adjust the system whenever feedback shows that system objectives are not being met. Persons using output should have some formal method of reporting conditions or levels of performance which they believe should be corrected. In some cases, special forms can be used.

Feedback is basically used for the following three reasons:

1. To adjust the processing
2. To analyze the quality of output
3. To predict the future quality of output

As feedback is obtained, processing can be adjusted to meet the objectives of the system. As errors are detected, the quality of output can be improved by their correction. Also, feedback allows systems users to predict characteristics of future output. Several years of correcting errors detected by feedback will assure greater accuracy in the future.

SYSTEM AND APPLICATION OUTPUT

Another way of categorizing output is (1) application output and (2) system output. Application output is obtained by programming the computer system. System output is programmed into the system by the manufacturer, such as error messages, communications with the computer operator, and other types of utility output.

APPLICATION OUTPUT

Basic types of application output are the following:

1. Printer page
2. Special preprinted forms
3. Microfilms
4. Plotters
5. Display (hard copy and screen)
6. Paper tape
7. Card
8. Mass storage, such as tape and disk
9. Audio
10. Telecommunications

Printer page: The type of output that can be obtained upon the printer page can be determined only by examining the characters available in the manufacturer's specifications. The systems designer should examine these characters before attempting to design any output forms for the printer page. Output on the printer page can use both upper and lower case characters in many systems. Although the printer page basically provides 120 print positions horizontally and 60 line spaces vertically, it can be adjusted (according to the model of printer used) for page length, carriage control, and paper size.

Special preprinted forms: Special preprinted forms come in many sizes and in almost any desired shape or color. Forms may be ordered with any printing upon it that is desired. Usually, a special form is used only when there is a need for large quantities.

Microfilm: Through the use of special hardware devices, computer systems have the ability to produce their output in microfilm form. This type of output is usually designated COM (Computer Output Microfilm) and is widely used and studied within the computer industry today. Microfilm output is many times faster than any other type of printed output and, of course, takes up less space than that needed on regular printed pages. Microfilm is basically used for storing historical data and for interchange of data where long distances are involved and mailing costs, freight costs, etc., have to be considered. With some of the microfilming techniques available today more information can be sent in a regular 8½ x 11" envelope than in an entire tractor-trailer truck using other means. The expansion of the use of microfilm is predicted in almost all areas. The systems designer is faced with a special problem when he must program and design microfilm output equipment into a system. The manufacturer's representative will be more than happy to acquaint the systems analyst with what he must consider when producing microfilm output.

Plotters: A drawing device known as a plotter can be used to produce output from a system. This hardware device can be directed by the computer system through the use of numerical digits and other methods. A drawing pen on the plotter moves across paper under control of the computer program. Maps, pictures of buildings, blueprints, and other types of drawing can be produced. The plotter is a useful device, although specialized. Most of the use of plotters is by engineering and scientific firms. If the systems designer foresees the use of a plotter within a system, the first step is to contact manufacturers and ask to be familiarized with the use of this device.

Display devices: Information can be displayed upon continuous paper forms similar to those used for ten-key adding machines or other types of continuous rolls such as those used in typewriters, teletype machines, etc. This type of display is usually called hard copy display. It is normally used for low-volume output and for interaction with the users of the system. This type of output may also come from type-writer terminals for use by the computer operator and from specialized devices which are used in remote locations.

Cathode-ray displays are televisionlike devices which display characters upon a television-type screen. These are used for inquiries into a system, usually a management information system or systems for hotel-motel reservations, etc. Cathode-ray tubes are advantageous because they can display hundreds of characters in a single flash on the screen. There is no need to wait for the information to print out.

Paper tape: Although paper tape is not used as much as it has been in the past, it is still a big factor in many systems which transmit data over communication lines. Paper tape is very similar to punched cards, except that it consists of a continuous roll of paper.

Cards: Output punch card devices are very common in all computer systems. Cards can be produced which are used as bills, invoices, and other types of documents. The basic use of card output is as a turn-around document. That is, the card will be produced as output from the system and will enter the system again at a later time as input. Another use of punch devices is to produce summary cards, which help to conserve use of cards. That is, if a company produces 50,000 cards a day, one summary card can be produced for those 50,000 cards. Such summary cards can be used at the end of the month to produce the monthly report. The system does not have to reuse all 50,000 cards produced each day to produce the monthly report. Use of summary cards should be considered by the systems designer wherever card volume becomes very great at the end of a cycle, such as the end of the week, end of the month, etc., and the organization does not use magnetic storage devices.

Mass storage devices: The most common mass storage devices are magnetic tape and magnetic disk, which are used to store data on a magnetic surface where it can be accessed by the system. There are many advantages and disadvantages to using both of these devices, and the systems designer should be familiar with all of them. For example, although magnetic disk is much more expensive than magnetic tape, it has some decided advantages not found in magnetic tape. The decision to store information on tape or disk is one which must usually be analyzed in detail by the systems designer.

Audio: Audio output of almost any kind is possible, but is only used at the present in specialized applications. The words or phrases desired for audio response can be recorded on some magnetic surface, usually a drum, and accessed directly by the computer to produce a string of spoken words. For example, the spoken words "one," "two," "three," etc., are recorded and reproduced as needed to form a spoken phrase such as "three, two, two, six, seven, eight, four."

Audio systems are used quite a bit in telephone information systems. A caller desiring information about a number may hear a recording which says, "The new number is 'six'—'five'—'two'—'seven'—'eight'—'four'—'three.' " Each of these voice recordings is selected by the computer based upon the digits recorded on a disk or some other medium. The difference, of course, is that instead of *printing out* a desired number, the computer *sends voice recordings* across telephone lines.

Telecommunications: Output to communications lines is now very common. Telecommunications lines can connect with other computers or with specialized terminals. This type of output is discussed in more detail in Chap. 12.

SYSTEM OUTPUT

Systems output, that is, those output messages that are programmed into the computer system by the computer manufacturer, can be categorized as follows:

1. Error messages
2. Systems logs
3. Operator communication
4. Status indicators
5. Dumps
6. Utility output

All these types of output give various indications as to how a system is operating. They also detect errors within the system. The systems designer does not need to plan for these types of output, as they originate from errors detected by the system itself. However, the systems designer should prepare written specifications as to what shall be done when error messages occur. Sometimes standard operating procedures can be given for each of the systems outputs; however, most manufacturers supply complete technical reference manuals about what output can be expected from the system and what action should be taken when this output is received.

Error messages: Error messages are those messages received when some error is detected by a system. These errors are not errors in logic but rather (1) errors in programming or (2) equipment errors. When an error message is received, the person concerned with such matters should turn to the appropriate manual and decide what action should be taken.

Systems log: As the computer is running, it can be set to log in the names of the computer runs, devices used by the system, etc. The systems log can be used for referencing runs, accounting information, audit trails, etc. Some type of retention period should be decided upon for systems logs.

Operator communications: Computer operators have the ability to communicate with the systems that they are operating. Messages will be received which indicate when the operator is to take action, and the operator can choose which type of direction he will give to the system. The alternatives are presented clearly in most technical reference manuals.

Status indicators: The status of a system is indicated by the computer. Examples of status indicators are indications of the amount of memory required by a program, which devices are available to the system, the time needed to run a program, etc. Most of these indicators are for use by the computer operator.

Utility dumps: A dump is any output onto a device from some other device or from the central processing unit. Examples of dumps are memory dumps, that is, the contents of the memory of the computer printed out on the printer, and tape dumps, which are the contents of the tape printed out on the printer. Dumps are usually used for checking contents of files as well as for checking program logic during the testing phase of programming.

Very rarely would a dump be included in a system application unless there is some need to check the accuracy of the data stored in the file. There are many better ways of making this check than dumping. However, dumps are sometimes used for audit and as control checks to prevent fraud and accidental misuse of files.

Utility output: Output from utility programs supplied by the manufacturer is another type of systems output. Utility programs basically relate to those common functions needed to run a system. Examples are sorts, merges, collates, and any type of data transfer. Utility outputs should be considered in a systems design. For example, the sort programs of most computer systems produce a record count of the number of records sorted. This record count can be compared to that of the next run to verify that the system is working correctly.

FORMS USE AND DESIGN

Several considerations must be given to the use of specialized or preprinted forms. First, the designer may have to design these forms not only for the computer but also for manual tasks which are performed within the organization. The basic first step in designing any form is to use qualified help or specialists in form design. It is not expected that every systems analyst and designer will be an expert in the use of forms and design of forms. Usually, the person utilized is the manufacturer's representative, that is, the salesman who sells the preprinted forms. Such persons are usually trained specialists in the use of forms.

The second step in designing forms is to obtain suggestions and final approval from all users of the form before it is sent for printing. Forms can be laid out in rough pencil sketches for this purpose. The value of forms ranges from the completely ridiculous to the very helpful. It is the job of the designer, of course, to eliminate the completely ridiculous forms and to improve upon the forms which are very helpful. Very few organizations can operate without certain types of forms.

One of the basic forms used by computer systems is the special preprinted continuous form. Certain considerations should be followed in designing these forms. The first consideration is whether a special form should be ordered at all or regular print-out paper used instead. Regular printer paper comes printed with black, green, or white lines, or as completely blank white paper. Sometimes the use of proper headings and other such output techniques will eliminate the need for special preprinted forms.

When special forms are ordered, the following matters should be considered:

1. Number of copies
2. Size
3. Coding words and phrases
4. Type of paper, paper weight, paper color, folding of paper, number of carbons, etc.
5. Possible use of the form as a turn-around document (will it allow mark sense and optical scanning techniques?)
6. Proper approval of the form

These considerations are very important to the systems designer. First, the number of copies which will be produced must be analyzed to make sure that all the copies are necessary. Following each copy from its origination as output to its final destination step-by-step can be a quick method of determining whether the form is necessary. Many persons in an organization receive copies of forms which go immediately to the waste-paper basket or which are filed and never used. Only the minimum number of copies of a form should be tolerated. However, if all copies of the form are to be used, as many copies as needed should be ordered, although most computers cannot print reasonably more than six to eight copies.

The next consideration is the size of the form. Will it file easily? Is there a standard filing size? Can it be mailed in a standard business envelope or does it require a special envelope? Size of the form will depend upon the type of application for which it is being used and how much information it must contain.

The third consideration is the codes and phrases to be used in the form. If codes are used, are they easily understood? Are they explained on the form? Are they nontechnical in nature? All words and phrases upon a form should be in simple English, as straightforward and brief as possible.

Another basic consideration is the paper of the form. How much does it weigh? How is it packed? How much storage is needed? How much will it cost to mail?

The color of the forms may be important. Form specialists can provide charts showing which color combinations are the most readable. The most readable colors are a combination of blue on white or black on white. The usefulness of forms is a more basic consideration than their aesthetic value, although they should be pleasant in appearance and act as a good public relations representative of an organization.

The final consideration is what type of carbon should be used. Some types of paper print multiple copies without carbon. This is a specialized type of paper and more expensive, but it may be easier to use. If carbons are used, the systems designer must remember to add to the systems flowchart the steps which will be necessary to separate the carbons from the forms. Normally, no output report is ever given to a user without first removing the carbons.

A final consideration discussed earlier is how the form can be used later as input. Forms must be designed carefully if they are to be used as mark sense forms or optical scanning forms. Examples of these types of forms are those used by credit card companies at the point of transaction, certain forms used for test grading, etc. The optical scanner (OCR) can read typewritten characters. Some machines can read only special type elements, but others can read regular elite or pica type from a standard office typewriter. The use of proper forms makes the optical reading of input much easier and less expensive.

A most important part of the design of forms is the final approval by all persons who will make use of them. The systems designer can carry the form around to all users to obtain comments, suggestions, and possibly written approval of its implementation. This may not always be necessary, but it will contribute to the acceptance of the form by the users of the system. A forms layout sheet can be used for the actual physical design, as shown in Fig. 6-11. This particular layout sheet contains a rough pencil sketch of a form. The rough sketch is turned into the final form by the forms manufacturer (Fig. 6-12). If the form is to be used as a source document, it should discourage longhand and encourage neat printing. Longhand can be discouraged by putting the areas for computer input data in character blocks.

The form should be self-instructing to persons using it, that is, each area of the form that is to be filled out should be self-explanatory. Usually, this requires a few words or phrases to be placed next to the area that is to be filled out. Use of square blocks is recommended, and the information as to what goes into this area should be within the block itself. For example, name, address, city, state, etc., should each be allowed one square block with the words "name," "address," "city," and "state" within the appropriate block.

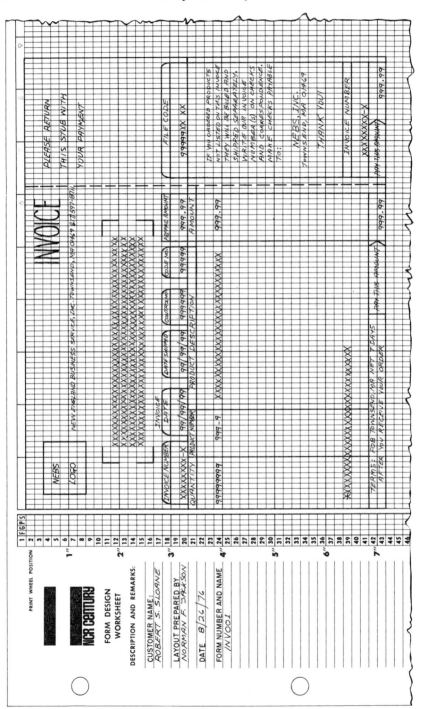

Fig. 6-11 Design of business form on a form layout sheet

Fig. 6-12 Finished business form of Fig. 6-11

EXAMPLES OF OUTPUT REPORTS

Figures 6-1 through 6-9 are examples of common output reports prepared by computers. Test yourself on your understanding of this chapter by answering the following questions about each of these reports.

1. Is this report (a) decision-making? (b) a requirement to meet a functional objective? (c) a requirement by law?
2. List the input items that you believe were necessary to produce this report.
3. Do you think this report meets the needs of its user? Suggest some improvements you might make.
4. Categorize this report as either (a) detail, (b) exception, (c) periodic, (d) management, (e) inquiry, or (f) information listing.
5. Suggest how control totals or record counts might be used to insure the accuracy of this report.
6. What key is used in producing this report?
7. In what sequence is the input file sorted?
8. Explain how you could apply a cross-foot check or a zero balance to check the accuracy of this report.
9. What editing (such as leading zero, dollar sign, etc.) is used in this report.
10. If possible, try to determine where limit checks could be used.
11. What items printed on this report were probably stored in the computer file medium as codes?
12. Does this report contain information that could be checked for accuracy by comparing it to past data?
13. Are there any flags in this report?

SUMMARY

Output techniques have been discussed before input techniques because inputs depend upon the type of output required. The output of a system is basically the meeting of the objectives of the system, that is, the objectives of the system are the output of the system.

Output is defined as anything which is produced by the system. Basically, output is used as a basis for decision making, to meet a functional objective, or as a requirement by law. Before output can be produced or designed, it must be analyzed as to what data is needed to produce it, what files it comes from, how it will be summarized, what is desired by the user, and other considerations. Although many types of outputs can be produced, the output of an information system is usually a report. Special reports are sometimes required, and inquiries are often made into a system for specialized information.

It is important to design reporting systems so that all reports do not come due at the same time. Therefore, the cycle of reports must be understood in detail.

Decisions that must be made about output before trying to design it include the reason for producing it, the data needed to produce it, its format, its volume, and its frequency. To facilitate the design of output, many record lay-out and printer lay-out charts are available. Output that is not used for reporting can be stored in some file medium. There is a difference to the design of report output and file output. To produce valid output, many techniques can be used. They include sequencing, cross-footing, control totals, record counts, reasonable tests, limits, historical checks, page numbering, use of headings, flags, etc. All are important and should be specified in written form by the systems analyst.

It is important for everything about output techniques to be documented for the programmer so that he may use them in producing his programs. Control over output is maintained by good written standards and written distribution rules. These controls are usually carried out by a section of the computer system called the data reduction section. Here, output reports are reproduced, bound, and distributed to the proper users.

Specialized types of output include real-time output (such as the output produced by reservations systems) and the output of other specialized immediate response systems.

The systems designer should make sure that systems do not produce an over-saturation of output, that is, that more output is produced than can be used. Since computers operate at high speeds, oversaturation sometimes occurs and should be checked. Feedback is a very important aspect of output. Output is monitored or measured, and this measurement is fed back to the persons operating the system. It is very important that feedback be acted upon as quickly as possible so that the processing and quality of the output of the system can be adjusted.

Many specialized hardware devices have been devised to produce output. The basic output of most computer systems is the printed page. However, microfilm, plotters, cathode ray tubes, mass storage devices, and telecommunications lines are all used as outputs. Besides regular application output, there is also systems output, which consists of messages and other communications programmed into a computer by its manufacturer.

Output forms must often be designed and printed especially for an organization. The basic rule is to work closely with the form manufacturer's representative before designing special forms. Things to be considered include the number of copies, the number of carbons, the size of the forms, the folds to be used, the kind of paper to be used, etc. It is especially important that all users of the form participate in its design and approve it as it will be printed. Two other rules of good form design are that the form should be self-instructing and should discourage long-hand by providing character blocks for the data required.

By carefully analyzing the output to be produced by a system in terms of the objectives of the system, a systems designer can be assured that his output will not only be useful but of considerable quality. The designer can then turn himself to the area of input and proceed to develop input specifications for the system.

Questions

1. Explain the three types of output.
2. How does a systems designer know that the output report to be produced will be accepted by the user? How does he know it is what the user desires?
3. Explain the difference between an exception report, a detail report, and an information listing.
4. Discuss the differences between output which is to be used and read by human beings and output which is to be stored in the computer for later use.
5. Assume that a data preparation section reports to the computer operator that 2,432 time cards have been keypunched. The computer operator runs the payroll

checks program starting at payroll check 1,642, prenumbered in increments of one. What should the number of the last payroll check printed be? What kind of output technique is this?

6. A manufacturer of portable typewriters knows that of all the typewriters produced, none sell for more than $500.00 or for less than $90.00; therefore, when invoices to customers are prepared for typewriter sales, all figures are checked before printing for less than multiples of $500.00 and greater than multiples of $90.00. What type of output technique is this?

7. The user of computer output reports complains to the systems manager that he cannot read or understand the report. The user makes the statement, "It is all Greek to me." Name and discuss the output techniques which might be used to correct this situation. Of course, this output report was programmed without any systems analysis or design effort.

7

Input

The analysis and design of input follows the analysis and design of output for two reasons. First, the input that will be used depends upon the output desired. Second, there is no need for any input that will not be used. Of course, if there is no immediate need for a particular data item, it might be included as input so that it will be available for some possible future use. However, basically the only input needed in a system is that which will be used to produce the output, that is, to meet the objectives of the system.

Careful analysis and definition of input are needed to promote efficiency within a system and especially to prevent errors in input data before it enters the system. It is the objective of this chapter to analyze all considerations involved in designing, collecting, and using data as input.

Input may be defined generally as anything which enters a system. In an information system, input is any character or group of characters which enter the system to be processed into output. Input characters are called data.

Input is the major problem of an information system. It is one of the most expensive phases of the operation of a computer system, and more errors and problems with the system can be traced to faulty input methods and input design than any other source. A most overused, but true, term in the computer industry is GIGO, which stands for "garbage in, garbage out." In other words, what is produced by a system can only be as good as what enters the system. Stated in more specific terms, the output report and other types of output from an information system can only be as accurate as the input which was used to produce the output reports.

The study of input design includes all phases of input from the creation of the initial data item to the time the input enters the system to be processed. Input not only can be created from scratch, it may also be received as the output of another system or some subsystem. The output of an inventory system may become the input to a cost accounting system. The systems analyst and designer must determine whether the input to be used by a system will be created and prepared by the organization, or whether the input can be received from some other system within the organization.

GENERAL INPUT DESIGN CONSIDERATIONS

As the systems designer approaches the problem of input design, there are several techniques which he may use. As already stated, he will want to use only those items which are necessary to the system (except for any items which may be possibly used in the future). As the analyst decides upon the items which are necessary, he also decides upon their form. Elimination of unnecessary characters and reduction of data items to compact and easy-to-use form without loss of information are initial considerations.

The analyst should attempt to understand in detail how the data is created and decide on his data "capturing" methods. The data format should be specified in detail and codes defined. As data items are specified, the designer should consider all outputs for which this input data is to be used. It must be remembered that data items may be used in many different subsystems within a system; in other words, various computer programs may use the same input data items.

To assure accuracy, checks and edit techniques must be set up to detect errors before they enter the system.

Based upon these general considerations, some definite steps to approaching input design can be stated.

STEPS TO INPUT DESIGN

The following steps are presented as an outline for the systems designer and analyst to consider as he approaches the problem of input design. These steps will be variously accomplished depending upon the problem which the designer faces. Some of them will overlap. However, if the systems analyst completes all the steps, he can be reasonably assured that his input has been designed in the most efficient manner and that errors will be detected. The steps are as follows:

1. Examine all output to determine what data items are needed.
2. Analyze future requirements.
3. Study the creation of data.
4. Determine procedures to transfer data from its place of creation to computer preparation sections.
5. Set time limits for data receiving and preparation.
6. Determine the methods of preparing the data.
7. State the characteristics of data items, records, and files.
8. Determine how data will be entered into the computer system.
9. Specify the edit and accuracy checks that will be used.
10. Specify batch control methods and checks on totals and record counts.
11. If data is by real-time entry method or remote terminal data collection, then controls should be specified, such as passwords and other security measures.
12. Prepare written specifications for all of the above steps.

These steps not only represent considerations which the systems analyst should think about as he approaches systems design, they can also serve as a check list when specifying what input will make up the system.

EXAMINATION OF OUTPUT

The first step is the examination of output to determine what data items are needed. Several methods are available. One used by many systems designers is to list all the output items in a column and then list all the input items which make up these output items. By cross-checking lists, every item needed to produce the output can be presented in a single list. When the data is specified at a later time, each item on the output list can be analyzed to make sure that its source is present in the input specification forms.

Figure 7-1 shows a simple list of output items in the left-hand column and the input items for this output at the right. Although some steps of input design can be accomplished as the output is being analyzed, it is usually true that no final input specifications can be made until the final file formats and output reports are agreed upon.

FUTURE REQUIREMENTS

The systems analyst will want to consider future requirements of a system, such as the needs of management within the next three to five years, and whether the system may eventually become part of a larger system, such as a total management information system. For example, if the system being designed is an inventory system, it is possible that it will eventually be integrated into the purchasing system, the cost accounting system, and the general asset system. Consideration should be given to what is required for these future requirements to be met without having to redesign the present system completely or change it too drastically. One of the most frustrating problems which a systems analyst or programmer can encounter is to need some data item that has not been provided for in the input design. Sometimes an entire reprogramming of the system is needed to accommodate the new data item. Also, nothing is more frustrating to a user of a system than to have computer personnel tell him that he cannot obtain a desired output because it has not been entered as input.

DATA CREATION

Data is constantly being created. Every purchase made by a consumer creates new data, such as number of units sold, reduction in inventory, credit amount, cash amount, etc. Within an organization data is created by purchase orders, supply requisitions, use of resources, hiring of workers, management decisions, etc.

When data is created by a transaction or a decision within an organization, there must be some way of capturing it for use in the information system. The most common method of capturing data is to write it on a form such as a sales ticket, a purchase order, etc. (Fig. 7-2). This written record of the data is called the source document. As mentioned previously, it is the systems designer's job to analyze all source documents used in a system and to redesign any that could be improved.

Several devices have been developed to capture data at its source. Employee badge readers may be used to clock employees in and out, thereby automatically capturing the number of hours worked by each. Cash registers have been developed which automatically record on punched paper tape the transactions that occur at sales stations. This paper tape can be used by a computer system to input the data into storage files. Light pens, typewriter terminals, OCR, ten-key adding machines, and many other devices have recently become available which not only record data as it is created but capture it in some machine readable form so that it may be entered into information systems without any human action.

Other decisions which the analyst must make are what types of codes should be used on the initial forms, how much data should be captured, what is really needed at the source of data creation, and how much time will be taken from a worker's regular job to prepare this input data.

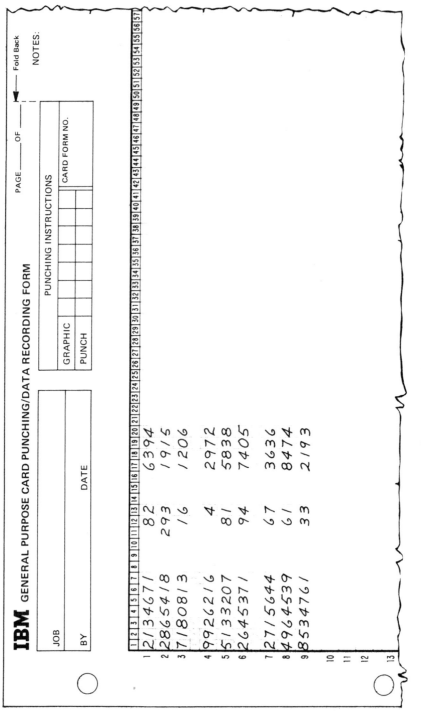

Fig. 7-1   Form for collection of data

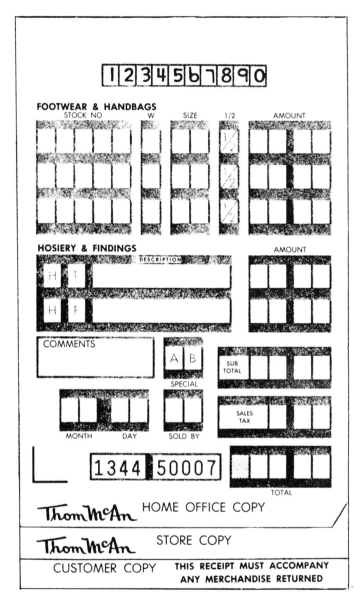

Fig. 7-2   Example of source document used to collect data from hand-printed figures

In summary, it is the systems analyst's job to determine the proper methods of capturing data as it is created. In some cases this may be done by machine; in other cases, it may be accomplished with pencil and paper and specialized forms.

TRANSFER OF DATA FOR COMPUTER PREPARATION

Once data has been created, it must be transferred from the place of creation to the place in the organization where it will be prepared for the computer. A systems flowchart should be drawn up to show all types of data movement. Usually, the department where data is created is responsible for gathering and making it available for transfer to the computer center, either through regular U.S. mail, the company mail, or hand carried by personnel from that department. For example, all purchase orders made by a purchasing department during the day can be transferred to the computer center at the end of the day. Usually, after the input data has been prepared, the forms are returned to the originating organization for storage.

Adequate "log" books should be used both to receive data as it comes into the computer data preparation center and to control sending and receiving of data forms.

TIME LIMITS FOR RECEIVING AND PREPARING DATA

Time limits are usually set by the systems analyst for the receiving of data from the originating organization. For example, if a payroll report is desired by management by Friday afternoon, then adequate time must be allowed for the preparation of data and the computer processing with a statement such as "All payroll time cards must be received by 5:00 P.M. Wednesday."

Expediting of data can be specified in input specifications as an aid to personnel within the data preparation section. Other types of time limits do not necessarily set a time for the data to be in, but rather are "cutoff" limits. For example, all sales orders received in the data preparation center by midnight of the last day of the month are prepared and become part of the monthly sales total, whereas those sales orders received after midnight at the end of the month become part of the next monthly report.

Essentially, data is set up to be received in cycles. A data item not received in one cycle becomes part of the next cycle. The cycles normally are daily, weekly, or monthly. For example, all inventory items used must be reported by 5:00 P.M. Those items received after 5:00 P.M. will become part of the next day's cycle. Commercial banks have made use of this method for many years. Banks may be open from 8:00 A.M. to 5:00 P.M., but only those checks and deposits received by 2:00 P.M. are processed with that day's transactions. Those items received after 2:00 P.M. become part of the next day's transactions.

DATA PREPARATION

As data is received by the data preparation section, it is usually assembled in batches. Each batch usually consists of that data received during one cycle of processing, but it may be divided into several smaller batches for control purposes. Each batch is usually assigned a number and a form called a batch control slip. The batch control slip states the number of the batch, the date it was received, the date it was prepared, and other such information. Totals for the batch can be calculated by use of a ten-key adding machine or some other type of device and are then recorded on the batch control slip. These totals can be used as a processing control later on. An example of a batch control amount might be the total hours worked by all factory workers during a payroll period. This number can be checked against future payroll runs to assure that all hours have been processed and follows the data through the system until the output is created. Later in this chapter accuracy of input is covered more thoroughly, and checking methods using batch control numbers are presented.

Sometimes data received by the data preparation section will be in a form which can be used for direct input to the system or which can be keyed by key operators into machine readable form (Fig. 7-3). Other times the data will not be in the form required by the computer system, and an intermediate step called coding is necessary. Coding is essentially reducing data to a more simplified form. For example, a personnel employment application form may state marital status as single, married, widowed, and divorced. This particular information would most likely be converted to numbers such as 1, 2, 3, and 4. It is more efficient if these types of items are coded before they are presented for data preparation. It is the job of the coder to transform such information onto a coding sheet which can be used for key operations. It is a well accepted rule within the computer industry that personnel preparing data by keypunching or some other key method should not have to think about the numbers and characters which they are punching. Each time they must stop to think about converting a word into a code, time is lost in preparing the data. The key operators should be able to keyboard exactly what they see in front of them and should not have to take time to convert codes or reduce data into abbreviated form.

Keypunching, that is, keying the characters from a source document to a punched card code, is the basic method of preparing data for the computer in today's industry. However, other methods are being used more and more. Keyboard devices are available which allow the operator to key the character directly onto magnetic tape or magnetic disk.

The systems designer should make sure that standards for keyboarding, such as how plus and minus fields are designated, what codes will be used to determine card numbers, etc., have been prepared by the data preparation section for reference by keyboard personnel. If these standards are not available, then they should be written by the systems analyst.

In most cases, a standard can be set for the amount of keypunching to be done by one operator. This has been found to be approximately 10,000 key strokes per hour. If the number of key strokes which the key operators are capable of is known, the time it will take to prepare data can be determined. A simple formula will quickly provide the analyst with a rough approximation of the time required. The formula is:

$$H = \frac{N}{\left(\dfrac{K}{R}\right)}$$

in which:

H = the number of hours needed to prepare the data
N = the number of records to be prepared
K = the key strokes per hour for each operator
R = the record length of the record

Assuming that the record length is 50 characters and that the key operator can perform 10,000 key strokes per hour, it will be seen that 200 records (10,000 divided by 50) can be prepared each hour. If there are 10,000 records to be prepared, it can be

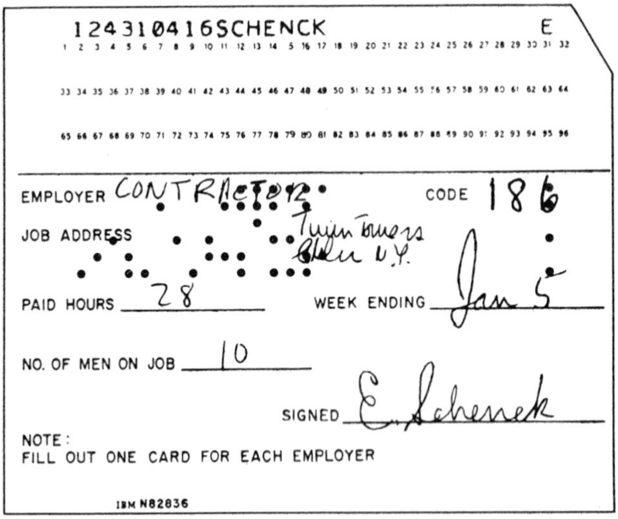

```
124310416SCHENCK                                    E
1  2  3  4  5  6  7  8  9  10 11  12 13 14  5 16 17 18 19 20 21 22 23 24 25 26 27 28 29 30 31 32

33 34 35 36 37 38 39 40 41 42 43 44 45 46 47 48 49 50 51 52 53 54 55 56 57 58 59 60 61 62 63 64

65 66 67 68 69 70 71 72 73 74 75 76 77 78 79 80 81 82 83 84 85 86 87 88 89 90 91 92 93 94 95 96
```

EMPLOYER CONTRACTOR: CODE 186

JOB ADDRESS   Twin Towers
                       Glen U.Y.

PAID HOURS   28   WEEK ENDING Jan 5

NO. OF MEN ON JOB   10

                 SIGNED E. Schenek

NOTE :
FILL OUT ONE CARD FOR EACH EMPLOYER

IBM N82836

Weekly workcard (actual size)

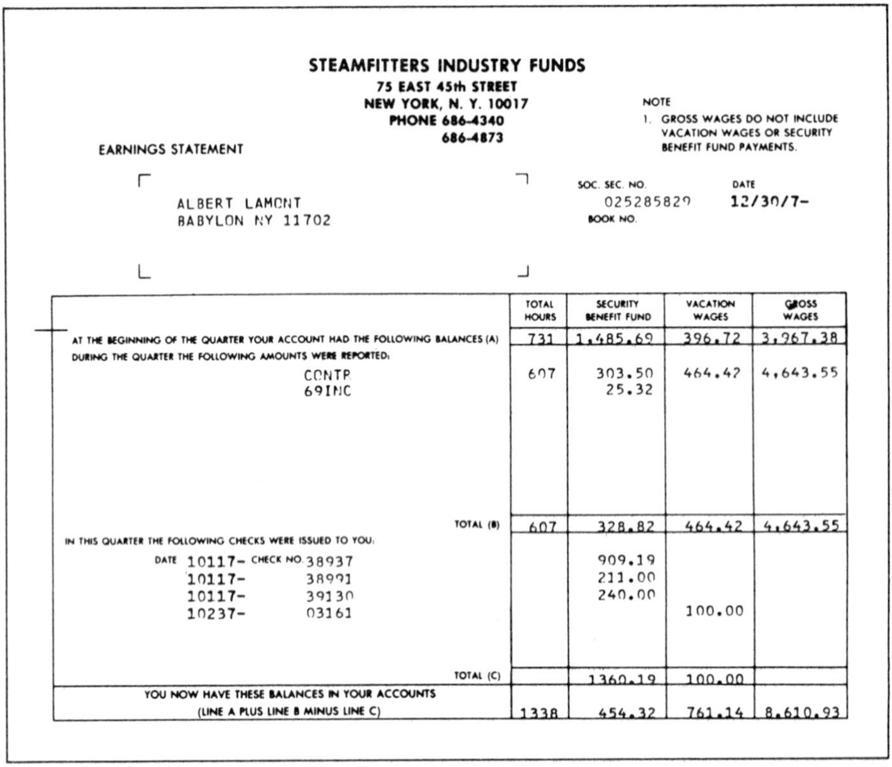

STEAMFITTERS INDUSTRY FUNDS
75 EAST 45th STREET
NEW YORK, N. Y. 10017
PHONE 686-4340
686-4873

EARNINGS STATEMENT

NOTE
1. GROSS WAGES DO NOT INCLUDE
VACATION WAGES OR SECURITY
BENEFIT FUND PAYMENTS.

ALBERT LAMONT
BABYLON NY 11702

SOC. SEC. NO.   DATE
025285829   12/30/7-
BOOK NO.

	TOTAL HOURS	SECURITY BENEFIT FUND	VACATION WAGES	GROSS WAGES
AT THE BEGINNING OF THE QUARTER YOUR ACCOUNT HAD THE FOLLOWING BALANCES (A)	731	1,485.69	396.72	3,967.38
DURING THE QUARTER THE FOLLOWING AMOUNTS WERE REPORTED:				
CONTR	607	303.50	464.42	4,643.55
69INC		25.32		
TOTAL (B)	607	328.82	464.42	4,643.55
IN THIS QUARTER THE FOLLOWING CHECKS WERE ISSUED TO YOU:				
DATE 10117- CHECK NO. 38937			909.19	
10117- 38991			211.00	
10117- 39130			240.00	
10237- 03161				100.00
TOTAL (C)		1360.19	100.00	
YOU NOW HAVE THESE BALANCES IN YOUR ACCOUNTS (LINE A PLUS LINE B MINUS LINE C)	1338	454.32	761.14	8,610.93

Fig. 7-3   Example of System 3 punch card and a statement produced from such cards

determined that 50 hours (10,000 divided by 200) will be needed to do so. In this manner, the analyst gains a rough idea of how much time will be needed for data preparation. By dividing the number of hours needed to produce the data by the number of key operators available, the number of hours can be reduced to the shortest amount of time. In other words, if 50 hours are needed to prepare the data and there are 10 key operators who can work on the data, then 5 hours will be needed for the entire section to prepare the data. This time should be doubled if verification is to take place.

Verification of data simply means preparing the data twice and comparing the results of the second preparation with those of the first. In a keypunching system, this is done by a card verifier. The keypunched card is placed into the key verifier and keyed again. If the two keyboard efforts are not exactly alike, a notch will be cut in the card which is punched the second time. The difference in the two cards can be analyzed and corrected by the keypunch operator. Although this is not a foolproof method, its accuracy is usually 99 percent error-free.

Besides keypunching and verifying, there are many other methods for preparing data for a computer, such as mark sensing, optical scanning, and the use of special data recorders, bar codes, specialized input devices, cash registers, badge readers, and the like. Where large volumes are encountered, it is most likely that the systems designer can use such methods to reduce costs. For example, large gasoline and oil companies use bar codes or optical recognition on transaction slips from credit cards. These characters can be read directly into the computer system and placed upon some computer file medium.

Where forms have to be typed, sometimes a special type font can be used that can be scanned by an optical scanner for direct input into a system.

Methods of using optical scanning, usually referred to as OCR (optical character recognition), are much simpler than for most types of input preparations, and regular typewriters, such as an IBM Selectric, can be used. The typist simply types the data on special forms or regular paper, and the sheets of paper are placed in the OCR machine. The data is read from the device by the use of light reflection and is placed upon some medium such as magnetic tape. Use of optical scanning has increased in the last few years and will continue to increase in the future. If data has to be typed anyway, typing it in optical scanning format can save much time by eliminating the need for keyboarding.

Selection of the type of input preparation is a decision which should be given special consideration. There is no easy formula for determining which method should be used. Such factors as equipment that the organization has on hand versus equipment that would have to be purchased are necessary considerations. Only if very large volumes are involved will the cost of optical scanning equipment and other such devices be economical. However, a basic guideline is to eliminate as many manual tasks in data preparation as possible.

Once data has been prepared by the data preparation section, it is again placed in batches and controlled by a batch control ticket. As the data is used, the batch control ticket can serve as a logging device and as a check on the accuracy of the system.

DATA ITEM SPECIFICATIONS

Before data can be prepared, there must be definite specifications as to what it consists of, its format, and its characteristics. Some of the basic considerations in specifying formats and characteristics are code compatibility, compaction, and the speed and efficiency with which the data can be managed.

Code compatibility basically means compatibility of codes between program runs and subsystems. In other words, if an inventory control system uses code 1 to mean parcel post, code 2 to mean air, and code 3 to mean freight, then the cost accounting system and other similar systems using inventory data files should use the same codes. It is only common sense that codes should be compatible within the entire organization.

Compaction of data essentially means the reduction of the space needed to store a data item and likewise the reduction of the time needed to prepare it. Compaction can take many forms. True compaction is the use of better coding schemes to represent more data. For example, two numeric digits can be packed into one eight-bit memory location. The achievement of true compaction depends upon the type of computer system being used. Other types of compaction might be called reduction; they include the abbreviation of names, the use of codes for words, and the use of certain characters to represent phrases or sentences.

Speed and efficiency in preparing input and in the actual use of input depend largely upon the data management system supplied by the manufacturer. This should be referred to when determining the input design specifications for any type of computer system. Basically, the data management system must always be considered the basis for specifying input for any computer model which an organization is using. As noted before, most data management systems differ from one computer to the next; therefore, the analyst must be familiar with the particular system for the computer he is using.

Two types of specifications must be made about input data. First is the specification for each input item or field. The next is the specification for each record or general specifications about the data file itself. These specifications are essentially the same as those noted in Chap. 5. However, they are repeated here in somewhat different format and in more detail.

INPUT ITEM SPECIFICATION

Each input item used by the system should include the following specifications (see Fig. 7-4):

1. Standard English name of item
2. Symbolic name of the item
3. How the data item is created
4. Method of preparing the item for the computer
5. The format of the item
6. When the data item will appear or not appear
7. How the data item is used
8. Forms upon which the item appears
9. Files in which the item will be stored
10. Action to take place if errors are found

Standard English name: The standard English name should be given to each item in the file. This is a name or phrase which fully describes the item. For example, if an item is an inventory control item such as unit cost, the name would read inventory unit cost.

English statement describing the item	How created	Symbolic name	Possible values	Encoding	Compaction	Type	Length	Use	Restrictions
Name of college department	By college dean	DEPT	Any approved curriculum	None	Truncate all but first four characters, i.e., English is ENGL	Alphanumeric	4 char.	Display	None
Inventory identification number	Assigned by Accounting Dept.	INVNM	100-100,000	None	None	Numeric	6 char.	Key or display	None
Sex	Applicant's form	SEX	1 or 2	Male=1 Female=2	None	Numeric	1 char.	Display	Must be transformed before display, i.e., 1 = male, etc.
Customer's balance	From computer program A33B	CUSBAL	Negative to 10,000	None	None	Numeric	7 num. 2 dec. places	Computation	None
Social Security Number	Application form	SOCSEC	0-99999999	None	Remove hyphens	Numeric	9 digits	Key or display	None

Fig. 7-4 Considerations of a systems analyst when defining data items

Symbolic name: The symbolic name is one that can be used within the system by programmers and others to denote the system in computer language. It differs from the standard name only in that it is usually abbreviated to conform to a particular manufacturer's computer system. For example, the inventory item unit cost might be named symbolically as INVCOST. Symbolic names should be as near to the standard English name as it is possible to make them and are preferable to symbols such as X, Y, and Z.

How the data item is created: Specifications as to how the item is created are necessary to understand where the item comes from and how it can be better controlled and analyzed.

Method of preparing the item for the computer: Usually, all data items within one record of the file will be prepared in the same way. In some cases, however, different methods of preparation are used, and the files are merged into one final file. The method of preparing the item for the computer, such as keypunching, key data recording, optical scanning, mark sensing, and the like, should be specified at this time.

Format: The format of the item refers to the item length, whether the item is alphabetic or numerical, and if numeric, how many decimal places it contains. Normally, input data formats are as simple as possible and contain no unnecessary characters. Dollar signs, commas, and other such symbols generally are not used.

When the data item will or will not appear: Sometimes certain data items are found only under certain conditions. The rules for when an item will or will not appear should be stated in the specifications. For example, in a transportation system, the data item on shipping "air" is to appear only when special handling is required. Where there is no item for air freight, normal shipping instructions are followed.

Many data items have rules as to whether they will or will not appear in the file. The systems analyst must also specify what will appear if the data item is not present. In other words, will the space in the record be filled by either blanks or zeros?

How the data item is used: A general narrative specification as to how the data item is used should be stated. Is it used in a calculation for some final total? Is it used to compute commission? All such questions should be answered at this time.

Forms upon which the item appears: All specialized source documents upon which the data item will appear should be stated. When more than one form contains the data item, specific designations as to which form will be used to gather the data should be made.

Files in which the item will be stored: Each place the data item is stored in the system should be named in this section of the specifications. Many data items appear in more than one file, especially key numbers, ID numbers, account numbers, and other such numbers.

How the item is edited or checked for accuracy: Detailed specifications on how to check and edit data items for accuracy appear below. Such specifications should be stated in this section of data-item specifications.

Action to take place if errors are found: The action to be taken on errors in data should be given. In some cases, errors are noted but the processing continues. In other cases, processing must be interrupted until the error is corrected.

All of the above specifications refer to individual input items usually called data items or data fields.

RECORD AND FILE SPECIFICATIONS

The following specifications should be given for each record or file within the system:

1. File name
2. How the file is created
3. Method of updating
4. Security
5. Priority
6. Storage media
7. Approximate number of records in the file
8. Sequence of the records in the file
9. Volume of the file
10. The frequency of use or appearance of the file

All these specifications must be made for the programmer, systems analyst, users, and managers to understand the system.

As the data items are gathered, they are placed in storage media known as files or data sets. There are many different files in a system. The systems designer must decide at this time whether to integrate the files into one large file called a data base or whether to maintain them as individual files, separated into logical divisions of data. For example, files that are separated into logical divisions, such as inventory files, sales files, accounting files, payroll files, personnel files, are usually considered an integrated data processing system. However, when an organization finds it is accessing data files frequently, it may place all its data in one large file, usually known as the data base or data bank. This is done only in organizations using a small number of data items or having a large amount of money to spend upon mass storage equipment.

At this time, the systems analyst will try to decide how the files will be created. Basically, for low-volume processing, where many different data items exist and many different types of runs must be made using the data, it is best to place data in individual files. For high-volume processing, however, where most of the data must be available for the output and where different types of data depend upon each other, it is preferable to use a data base. Most business information systems executed by computers use individual files. Data bases, on the other hand, are used to store large groups of data which can be built into one large common file; for example, the criminal records of the FBI or the license and vehicle registration of state governments. Even when an organization has a data base it will still have need for specialized files in other areas. The following paragraphs explain the specifications for files and for records within the file.

File name: The file is given a name which is used throughout the system for that file and which is used whether the file is input, output, updated, or some other type. The file name must conform with the particular computer system the organization is using. For example, most names cannot contain more than eight characters and must begin with an alphabetic character. File names need not denote the contents of the file; rather, they are codes which describe the file in few characters. Chapter 5 gave examples of such file names. For example, PAYR613 represents a file in the payroll system. The R means randomly accessed. The 6 means cost accounting department. The 1 could mean first-of-the-month run, and the 3 could be the priority over other runs. This is the way files should be coded so that the file name explains not only what the file contains, but how it is used. As explained in Chap. 5, this coding scheme should be standard for all computer use.

How the file is created: This is essentially the same specification that is given for each individual data item. However, where the data items are created by different methods, the file itself may be created by merging or collating other files. If a file is created by merging, then it should be stated at this time.

Methods of update: Files never remain static unless they are for historical records only. Data in files changes constantly: addresses, phone numbers, and, of course, amounts and other accounting figures. A specific narrative should describe the methods used for updating. If the file being specified is a master file rather than a transaction file, then the method of creating the transaction file should be stated.

Security: Security of a file—whether it is confidential, secret, open, or other—should be stated. Also, if appropriate, the physical security to be taken with the data file can also be stated, such as, "Store each night in a locked safe."

Priority: Priority of processing a file or securing a file should be stated. In some instances, priority numbers can be given to the file; in other cases, a general statement of the priority of the file is adequate.

Storage media: Data may be stored in many different forms—cards, magnetic tape, disks are all examples of data storage media. The method of storing the file should be given, as well as some information as to amount of storage that will be needed and possibly an indication of the cost of storing the file.

Approximate number of records in the file: The number of records in a file will not always be known; however, some approximate number usually can be given. For example, an organization generally does not know the exact number of items in inventory at any one time, but it will know the approximate number. Specifying the approximate number serves as a check for processing of the file. For example, if approximately 10,000 records for inventory are in the inventory file and the computer system counts 20,000, then some indication of error is available.

Sequence: The sequence of the file, whether ascending or descending, and the key numbers or key identifiers upon which the sequence depends should be stated.

Volume: The volume of the file is stated in the sense of processing and updating the file. In other words, how many transactions affect the file during each processing run? This is an approximate number but should be given for control purposes.

Frequency: The frequency of use of the file is stated as the cycle of its use: daily, weekly, or monthly, or some other frequency? In some cases, frequency may not be known, but usually file processing is performed in cycles.

METHODS OF SPECIFYING INPUT

The most common way of specifying input is by the use of layout forms. Layout forms are usually prepared and obtained from the computer manufacturer. A layout form contains both the space and designations needed to specify file or input design.

Card design: After determining input data specifications, the systems designer must next consider the card design. How data items are placed in the card or the order used is not critical. However, items should be together with no blank space between them. With punched cards, the use of a second or third card should be made only if necessary. If second or third cards are used, the ID number of the record should be repeated in the second or third card. When using punched cards, the systems designer can either use a standard card or obtain preprinted cards. Preprinted cards are designed by the systems designer and obtained from the punched-card manufacturer. A rough diagram of how the card should be formatted is given the manufacturer and a final mat for approval is prepared before the cards are printed. Most systems designers commonly use a multicard layout form. These card layout forms can be passed on to the programmer for specific designation of the input of the system. Black diamonds, called carets, are used on layout forms to denote decimal position (see Fig. 7-5). Punched cards usually cost less than three dollars per box of 2,000.

Record layout forms are used for both magnetic tape and magnetic disk files, two of the most common types of storage media in use today. Usually, on both card and record layout forms a line is drawn to separate data items; the name of the item is written between the lines. Whether the field is numeric, alphabetic, or alphanumeric can also be denoted within the lines.

In some cases, special layout forms—such as those used for mark sense, bar codes, and optical scanning—may be needed. Such forms should be obtained from the manufacturer of the device. The principles for using special forms are the same as those for all other layout forms.

Use of optical scanning methods: Optical scanning, known as OCR (Optical Character Recognition), is being used more and more to prepare input data. Almost any typed or printed character can be scanned and entered into some type of computer media by equipment. Usually, the only considerations are the cost of obtaining the equipment and the system changes necessary when optical scanning is introduced.

Fields of data items within the records are separated by slashes. This permits variable length when preparing data items, which are expanded into fixed lengths by the OCR system. Once the machine has scanned the records, the data is then made available on magnetic tape or magnetic disk. If desired, the computer can further edit the data and place it in new format.

Optical scanning methods will probably produce more errors than punched card verification methods; however, some data such as mailing lists do not require critical accuracy. Therefore, optical scanning can reduce costs as well as prepare the data adequately. A significant advantage of optical scanning is that data can be prepared on a regular typewriter and no special equipment such as keypunches, key verifiers, etc., are needed.

ACCURACY OF INPUT

A computer printed out a stock dividend check for $784,000.00 and the check was mailed. It should have read $78.40.

A farmer in the Midwest one day received 37,000 identical letters asking him to subscribe to a magazine. The mailing labels had been printed by a computer. (The farmer is reported to have written in reply, "I give up. Send me a subscription.")

Customers of a large credit card system were sent a notice that said they must pay the above balance immediately or legal action would be taken. The balance read $0.00. (One customer found the only way to stop the computer system from sending him this notice was to mail in a check for $0.00. Evidently, the check was processed, the account updated, and the computer reported the account paid in full.)

All the above computer errors and many others have been reported. Although

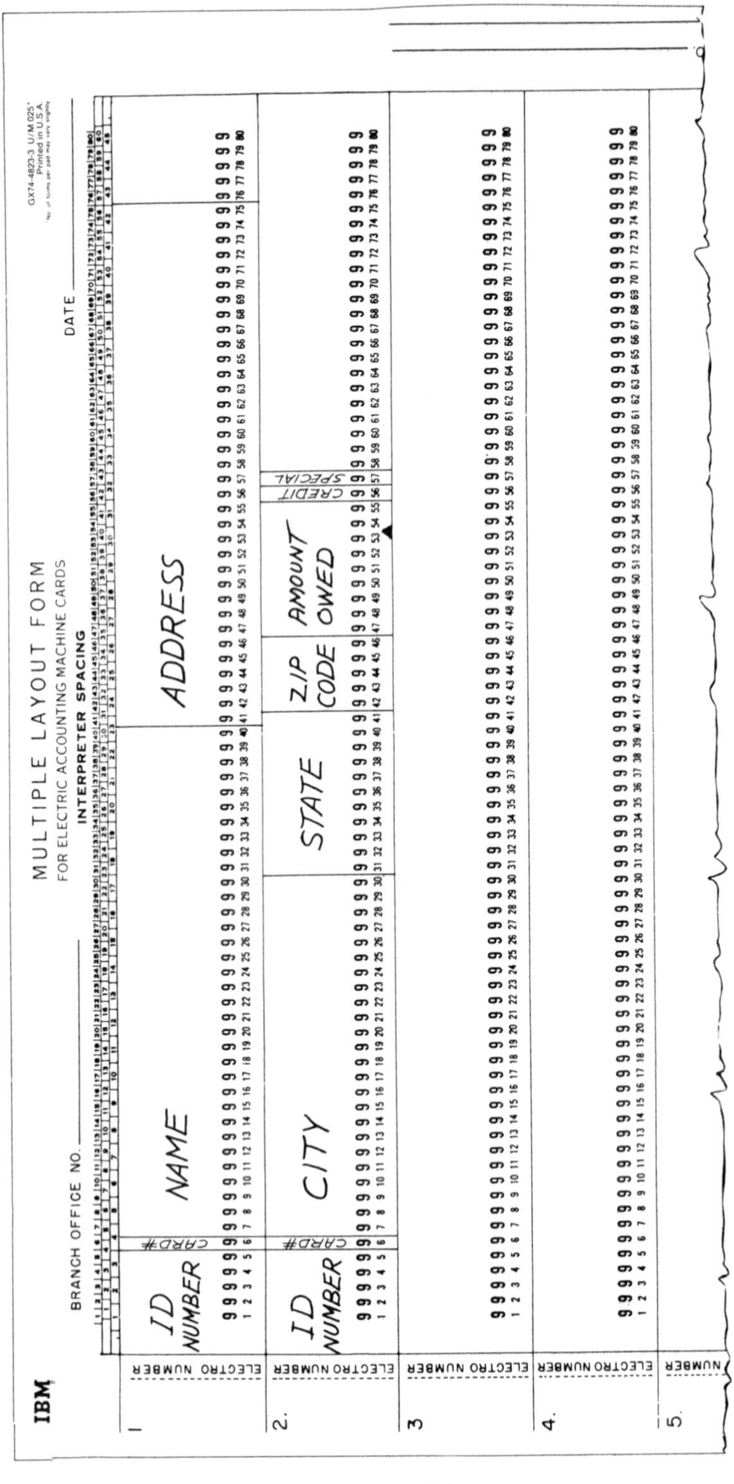

Fig. 7-5 Input card design

some errors are attributed to processing logic and output, probably most errors are caused by the inaccuracy of input.

There are three basic methods for assuring accuracy of the input items which enter a system. The first is to verify the data as it is prepared. Since data has to be keyed twice, this doubles the cost of preparing the data. One way to reduce the cost of this method is to verify only that input which is critical or that data which samples or other techniques may have shown to be in error. Also, the performance of key operators can influence a decision as to whether to verify data. Experienced operators who have established a record for accuracy may require fewer data verifications than new or inexperienced operators. Again, certain data, such as mass market mailing lists, does not necessitate character by character verification. In other cases, such as accounting figures, it is very important that the data be exactly as shown on the source document.

A second way to assure data accuracy is simply to review it physically. Statistical sampling techniques can be used to determine what will be reviewed. Certain computer personnel can be designated to review the data before it enters the computer system. Data can be reviewed manually; for example, every fifth card which is punched can be checked against the source document to determine whether the data is being punched correctly and to catch whatever errors are possible. However, the fewer manual steps used in data verification, the lower the cost of preparing input.

A third way to verify data is to use editing techniques. These techniques are developed from past experience and should be applied to all data items when possible. Most data items have certain characteristics which a computer can check rapidly and accurately. Figure 7-6 shows an edit program, which is the first check on data. The edit program checks the known characteristics of each data item. Data containing error is returned to the entry point of the system to be corrected. Once data is edited successfully, it is then summarized and control totals are checked against batch control tickets. If the number on the batch control ticket is not the same as the control total produced by the computer, then an error is indicated, and the data is returned to be checked by the data preparation section before it is used in the system. These two

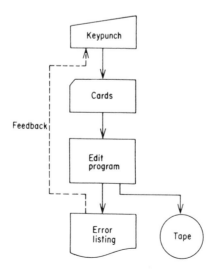

Fig. 7-6   The edit program

checks assure that inaccurate data will not enter the system. The systems analyst and designer should use these methods at all points where data initially enters the system. Many of the errors that appear in systems could be prevented by simple batch control and edit techniques.

INPUT EDITING TECHNIQUES

The list below enumerates the different techniques that can be applied to determine the quality or accuracy of data items. These techniques may be applied more than once to one particular data item.

1. Type of data check
2. Validity check
3. Record count
4. Association tests
5. Check digits
6. Algebraic signs
7. Completeness test
8. Hash totals
9. Removal of unneeded data
10. Sequence checks
11. Batch total
12. Cross-check against master file
13. Limit or parameter checks
14. Arithmetic checks
15. Table comparison
16. Modulo or residue checks
17. Reasonableness or historical checks

Type of data check: The first basic data check that should be made is to determine whether the data is the type expected. For example, a Social Security number should be all numeric; a name field normally will be all alphabetic. Such checks are simple operations for most computers.

Validity checks: The second basic data test is a validity test. Certain codes and values of data are known; if a certain data field should always be a one or a two, the appearance of any value other than a one or a two means the data is invalid. These checks also are simple unless many different valid codes exist. In that case, a different type of check is usually needed.

Record count: The basic method of controlling and determining whether data has been processed correctly is the record count. If the records in the file are counted as they enter the system, probably at the time of edit, the count can be carried with the file throughout the processing of the system. In fact, record counts can be placed magnetically in the trailer label of the file. For example, if there are 10,461 records in the inventory file when it is edited for input, then there should be 10,461 records in the file when it is processed by other programs. Record counts are similar to batch controls in that they assure that all records have been processed and accounted for.

Association test: An association test determines if the data is logically associated with other data items in the record. For example, if a warehouse for a company carries only certain items, that particular warehouse number must appear when the item number appears. By using association tests whenever possible, data can be cross-referenced and cross-checked.

Check digits: A check digit is a digit either appended to the normal data item or created by a manipulation of the data item. A check digit prevents fraud and also detects keypunch and other types of input errors in the system. It should be specified by the systems designer for control purposes.

A very simple check digit may be created by appending a number to every account number. During input edits, every account number could be checked to see if it contained this check digit in the one's position. A check digit may also be the sum of all the digits in the number. For instance, the sum of all the digits in the number 417 is 12. The 2 of the 12 could become a check digit used to check for accuracy. However, this method can miss certain types of errors. A transposing of 147 for 417 would not be detected, because the totals for the numbers are the same.

Many credit card numbers contain a check digit obtained by manipulating the account number. For example, dividing the number by six, multiplying it by the square root of that number, and adding twelve to it would create a simple check digit which could be appended to every account number. If the credit card number which enters the system does not have the appropriate check digit after this manipulation, then either an error in input or perhaps fraud is evident. Other check digits are created by casting out certain digits. For example, a well-known system of casting out nines can be used to create a check digit which is the total of the numbers after all nines have been cast out. In another cast-out system, every other digit in the value can be eliminated, and the check digit is simply the sum of every other digit in the value.

There are many types of check digits and the systems analyst should analyze carefully the check digit method he wishes to use. In some cases, he may create his own method. It is probably best to create a check digit based upon intricate manipulation of the numbers—such as dividing, subtracting, multiplying, adding, using the square root—and the addition of a constant number. This prevents a person outside the organization from deciphering the check digit and attempting to counterfeit a credit card, since it would be nearly impossible for him to know what method was used to create the check digit.

Algebraic sign checks: Checking signs of a field is a simple way to detect errors, although sometimes inappropriate. Certain types of data will always be positive, whereas other types of data will always be negative. When this condition exists, the item should be checked to see if it is negative or positive.

Completeness test: A check to see that data is present should be made on all input data. Computer systems can check easily for blank spaces or zero conditions in data item fields. In some cases, it is appropriate for the data item to be missing; however, in most cases where the data item is essential, checking to see if the space where the data item should be is either blank or zero will reveal many errors.

Hash totals: Hash totals are totals of numbers not normally used in calculations. For example, a total of all Social Security numbers used in a personnel file would be a hash total. These hash totals can be used to check completeness of processing. For example, if a hash total of all Social Security numbers is taken during input editing, then this hash total can be checked during processing to see that all items have been processed. If hash totals do not match, then all items may not have been processed or the program logic may be incorrect.

Removal of unneeded data: The removal of all unneeded data characters should be specified in the input edit design. Characters such as dollar signs, hyphens, and commas are usually not needed in input numeric items. They can be eliminated. Removal of all unnecessary data characters will considerably improve system efficiency.

Sequence check: In all cases where data is in some form of sequence, a sequence check should be made. An ascending check determines that each sequence item is greater than the previous item according to the computer collating sequence. Items out of sequence can be easily checked and flagged for correction.

Batch control totals: Batch control totals are an important technique of input editing. They generally are taken manually, using an adding machine, and are written on the batch control ticket which accompanies the data file. Checking batch control items can assure that all items within the system have been processed. The biggest drawback and disadvantage of batch control is that batch control totals must be calculated manually, which takes time and, therefore, is costly.

Cross-check of input against master files: Data entering the systems, which is transaction data, can be cross-checked against the master file. For example, identification numbers can be checked against the master file to see that each identification number is part of the system. The systems designer can specify this check as part of the input edit run.

Limit or parameter check: An important check, the limit or parameter check sees that the value of a number does not exceed a limit or fall outside of a parameter. In a system using the codes 1, 2, 3, 4, 5, the code should be checked to see that it does not exceed the value of 5 or is less than 1. Sometimes data items always fall within certain parameters. For example, a company may sell products which cost at least $10 but not more than $20. An amount less than $10 or more than $20 per item would be checked for possible error. Limit and parameter checks should be used freely and specified in the system where the limit could not possibly be exceeded. There are several instances of computers printing million-dollar payroll checks and insurance refunds when the actual figure should have been much less. A simple limit check might have prevented this. The limit or parameter check is more of an output technique, but it has many applications in input checking also.

Arithmetic checks: Certain arithmetic checks can be made against data. In some cases, data item A plus data item B should equal data item C, or 25 percent of A plus 75 percent of B should equal 100 percent of C. These checks can be made when the data is known to have some mathematical relationship. Other types of arithmetic checks are checks to see that one value is greater than another value and vice versa or that the values are always equal.

Table comparisons: A table comparison is basically a check against a table of values to find a valid value or code. For example, if the input data item is known always to be one of ten different percentages, then the input item can be checked against the ten different percentages in a table stored in the computer. Table comparisons can be used to check tax rates, insurance rates, and alphabetic codes.

Modulo or residue check: A residue or a modulo check is one that uses the results of some calculation. It is similar to a check digit. For example, any number divided by 4 will have a remainder of 0, 1, 2, or 3. Therefore, identification numbers assigned to keys such as ID numbers, account numbers, bank numbers, and code numbers can be specified to be those numbers which have a residue or remainder of 2 when divided by 4. Or, every time the identification number is divided by 4, the remainder should always be 2. These numbers can be calculated by the computer and used to assign account numbers and other identification numbers. As they appear in input transactions, the residue check can be made to determine if it is a valid account number. Other types of residue checks are Modulo 9, Modulo 7, Modulo 11, etc., where the residue or remainder of a value is determined by division or some other manipulation of the value.

Reasonableness or historical check: A reasonableness check is very similar to the limit or parameter check. However, in some cases, a reasonableness or historical check is preferable. For example, if experience shows that the average amount of sales never exceeds $300, this limit can be set in the machine as a reasonable high limit for all sales orders created as input. Violations of reasonable numbers, parameters, and limits do not necessarily mean that the data is inaccurate, only that it may need special attention. This check is also called the historical check because historical figures can be used to determine what is reasonable and unreasonable. If a company has consistently found that its employees average six overtime hours each week, then using the number 6 to check the average number of overtime hours is a good historical check upon accuracy of calculations and input into the system: The total number of overtime hours divided by the total number of employees would normally be approximate to six.

All of the above checking techniques should be applied as frequently as possible to assure accuracy; however, they must be included in the systems design specifications or the programmer may ignore them. Many errors can be avoided simply by editing the input data to assure its accuracy. Although errors will always creep into any system, their detection is possible and an input editing check can be included in any systems design (Fig. 7-7).

DOCUMENTATION

All the specifications covered above should be written into a package which will be part of the systems specifications documentation. Through written specifications, the systems analyst communicates with the personnel who operate the system, particularly the programmers. Further details on documentation are included in Chap. 10. Once output and input specifications are completed, the systems analyst can develop the procedures and logic necessary to process the data. However, the written specifications of formats should be prepared and distributed before work begins on processing specifications.

SPECIAL TYPES OF INPUT

There are many special types of input devices, among them badge readers, credit card forms, mark sense, bar graphs, department store tags, light pens, and the like. Also, there are recently developed systems that use remote terminals to capture input, such as cash registers connected by telephone lines to a computer system. It is possible that, some day in our society, every transaction will be immediately entered into the computer. The analyst should familiarize himself with many of the special input devices, for they can save time and improve accuracy.

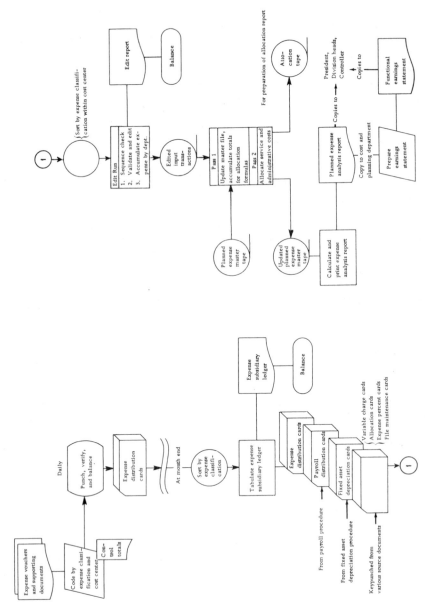

Fig. 7-7   Flowchart showing where the editing takes place in a system to prepare expense reports for a bank

DIRECT INPUT VIA TERMINALS

An increasingly utilized method of obtaining input for computer systems is through remote terminals. These are varied and operate in many different ways. The systems analyst may use all the techniques available to him in normal integrated computer processing—such as reasonableness, limit checks, check digits, etc.—but he must have some way to control input into the remote terminals. In one method, data is captured into the system and immediately stored upon some file medium. Then, although collected by remote terminal, the data is treated as if it were prepared by key-punching or any other method. A basic control that should be mentioned here is the control of "logging." As input comes from terminals, there should be some type of log available which tells when the data was transmitted and what its volume was. Although logging is not always possible, the analyst should consider it.

SUMMARY

Input has been defined as anything which enters the system. Normally, in an information system, this is called data and consists of raw characters of both numeric and alphabetic form. Before input can be specified, output should be analyzed and designed, for input depends upon the system objectives, or output. Also, the quality of output depends directly upon the quality of input.

Some general steps which can be followed in input design are (1) examining output and determining what data items are needed, (2) analyzing future requirements, (3) studying and listing places where data is created, (4) determining procedures to move data from creation to computer center, (5) setting time limits for receiving data, (6) determining methods of preparation, (7) stating the characteristics of the data items, (8) determining how the data will be entered into the computer system, (9) determining the file media on which it will reside, (10) specifying edit and accuracy checks, (11) specifying batch control methods and total checks, and (12) preparing written specifications for all of these steps.

The analyst is concerned with specifications of input data from the initial point of creation until its entrance into the system. The systems analyst should specify in detail the method of preparing data and moving it from point to point.

Card layout forms, record layout forms, and other special forms are recommended in preparing input specifications; additionally, written specifications which detail each data item and information about each record and file should be prepared. Data should be thoroughly checked (edited) before it is entered for use in processing by the system. The editing techniques specified in this chapter are not complex, and in most cases, involve very simple computer operations; however, they correct many problems.

There are many special devices which can be used as input. Many of these, such as remote terminals and remote collection devices, are being used much more today; and the trend is to use more specialized types of input capture.

Questions

1. Why does the analysis of input come after the analysis of output?
2. Name three general considerations of input design.
3. How is input data determined?
4. What is meant by data creation?

5. Name some methods of transferring data from creation point to the computer center and how this can be controlled.
6. Explain batch control.
7. What is a source document?
8. Explain the use of layout forms.
9. Name and describe four methods for preparing data for a computer.
10. Tell why the accuracy of input is important and how this importance relates to other phases of systems analysis and design.
11. What is OCR?
12. List and describe five methods of checking accuracy of input.
13. When should input editing be used? What happens when errors are found by editing?
14. What is the scope of input design? That is, where does it start and where does it end?

Problems

1. Draw a systems flowchart of input preparation from data creation to storage of the data on some file medium.
2. Draw a layout of the input that would be needed to produce your bank checking account statement.
3. Figure the approximate time that would be needed to prepare 100,000 records of data where the record is 78 characters long and the keyboarders average 6500 keystrokes per hour.
4. Prepare the input item specifications for the data item "amount owed to vendor." The data item is from purchase orders for structural steel.
5. Write the documentation that would be needed for a student file in a university.
6. Assume that a cost accounting report uses four columns of figures from input data to produce a fifth column. The final total of the fifth column is compared to the projected total for the month. Prepare specifications of what input editing techniques could be used to assure accuracy. There are never more than 600 input items.
7. Design a department store bill. Now try to determine how the data would be collected to produce the bill. What various methods of input capture could be used?

8
Procedures and Logic

The modern computer system operates by sets of step-by-step instructions called programs, which are executed by the central processing unit. The systems analyst develops the logic and special procedures for the computer system; he also specifies manual tasks for programming personnel to carry out. The detailed programming of the system is left to programmers.

In some organizations, programming is carried out by the systems analyst, who is known as a programmer/analyst; however, in most organizations, systems analysis and programming are separate. The systems analyst carries out the analysis and design phase of systems development, whereas the programmer actually writes the instructions for the computer system and implements the program runs on the computer. This chapter will survey the considerations and tasks of the programmer and analyze those parts of programming where the systems analyst may participate.

THE PROGRAMMING CYCLE

As the programming phase begins, the systems team turns over to the programming section all the input-output specifications and other written documentation about the system. The programmer should have access to all systems documentation, although he usually does not require management summaries, considerations as to selection of equipment, and other such documents. The programmer's concern is to make the system work. To accomplish his task, he usually performs six different steps. These steps are usually known as the programming cycle:

1. Problem analysis
2. Flowcharting
3. Coding
4. Testing
5. Implementation and operation
6. Documentation

The programmer performs these steps on one program run at a time within the system. In a systems flowchart, the rectangular symbol denotes the processing done by the computer system.

Problem analysis: Before a programmer begins writing a program, he should first understand the over-all objectives of the system and what the program should accomplish. Analyzing the program may also require research and some

137

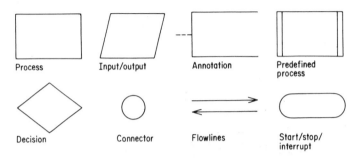

Fig. 8-1   Basic program or logic flowcharting symbols

interviewing. For example, a programmer may write a cost accounting program, but he may not be a cost accountant. If he is not, then his job might include becoming familiar with cost accounting principles and their use in his particular program. Programmers are not expected to know the details of all the application areas in which they work. Therefore, the detail of the logic sometimes is developed by the systems analyst or a specialist in the area.

Although the programmer is not expected to be an expert, his job during the problem analysis phase of the programming cycle includes finding out the principles involved in a discipline by library research, interviewing, or instructions from the systems analyst and designers. Also, during problem analysis, the programmer usually familiarizes himself with the input data items used in his program. Sample forms, systems flowcharts, print charts, and other aids will help him better understand how the program should work.

Flowcharting: The programmer prepares a flowchart very similar to the systems flowchart prepared by the systems analyst. However, program or logic flowcharting depicts much more detail and uses different symbols. Logic flowcharting uses the four basic symbols shown in Fig. 8-1: the processing symbol, the decision symbol, the connector, and the input-output symbol.

The rectangular processing symbol denotes any computer processing operation such as add, subtract, multiply, divide, move, etc.

The diamond symbol denotes a decision or a change of sequence in the program based upon some condition. For example, in a hotel reservation system, there would be different processing for "room reserved" and "no room available." Therefore, at some point in the program, the computer must check the number of rooms available to determine whether a room could be reserved. This particular point in a computer run is a decision point. The decision symbol usually contains one flowline entering the box and two or more flowlines leaving it, therefore denoting a change in sequence.

The input-output symbol is used where data is transferred in and out of the system. A record transferred from magnetic-tape storage into the computer's memory would be denoted by the input-output symbol.

Connector symbols are used to denote the next step in the program. Sometimes connector symbols are used to show the continuation to another page or to a subroutine or some subfunction within the program.

Flowlines are used to connect symbols.

Other symbols sometimes used in logic flowcharts are annotations, which denote the changing of the logic or a comment about the program, and the predefined process symbol, which denotes that some operation or some sequence of program steps has already been specified in a subroutine or in some other portion of the documentation. The terminal symbol can denote a stopping point, a halt, or an interrupt within the system.

The programmer uses these basic symbols to develop the step-by-step logic of the computer program. He may use detailed computer language statements within his symbols to denote what process is being carried out, or he may use English phrases to make the flowchart more readable. The flowchart provides both proof for the programmer that the program will work as he envisions it and a communication tool for interaction and discussion about the program run. The logic flowchart also becomes a basic part of program communication.

Coding: Once a detailed flowchart is drawn, the programmer can begin coding. To code a program, some particular computer language must be selected. In some cases, the systems analyst will specify a standard language to be used throughout the system, but most likely installation standards will specify to the programmer what language to use. The most common languages used today are COBOL, FORTRAN, RPG, and the assembly languages for the particular model computer.

COBOL (Common Business Oriented Language) is usually the language selected for business and commercial applications, such as banking, insurance, manufacturing, and accounting systems. An English-type language, COBOL is widely accepted for programming integrated data processing systems. Probably 70 to 80 percent of all the commercial systems operating today have been programmed in COBOL.

FORTRAN (Formula Translation) is a mathematically oriented language which can be used for specialized problems in engineering or detailed mathematical calculations.

RPG (Report Program Generator) is a specialized language used mostly on small computer systems for business type processing.

These three languages—RPG, COBOL, and FORTRAN—are used in most programs written for computer systems today. PL/1, Basic, and other such languages are used in some specialized cases; however, they are not widely accepted for systems work.

Assembly language is one which is used in a system that requires specialized operations. For example, a system which uses telecommunications lines, cathode-ray tubes, badge readers, and other such special devices normally cannot be programmed in any of the higher level languages such as COBOL and FORTRAN. Therefore, assembly languages (the computer manufacturer's language for a particular computer) must be specified to be used for programming.

All computer systems have their own code, which is essentially a binary language which uses binary states for operation. A translation is necessary before COBOL, FORTRAN, or any other language can be used within the system. Figure 8-2 shows the steps the computer system will normally carry out to create a program from codes of FORTRAN, COBOL, and other high-level languages. These languages are usually called the source language; the machine language produced by the system is called the object language.

Testing: Testing of a program—sometimes called debugging—is simply running the program and analyzing the results. Every conceivable program operation should be tested—that is, every path of the program and every type of data used—to see that the program works correctly. One testing method is to prepare data which tests

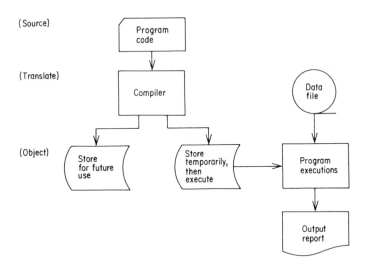

Fig. 8-2   Simplified view of source language—COBOL, FORTRAN, etc.—used as input to the compiler and producing object code, or machine language (this process is to be assumed by the Systems Analyst and not used as part of his flowchart)

all the different tasks of the program. The results of this run can be precomputed manually and compared with the actual data or information produced by the computer.

Implementation and operation: The programmer is responsible for seeing that his program integrates into and works within the system. Once the program has been tested, however, it is turned over to the computer operators, who have the responsibility of running the computer system. During systems design, some specialized operations, such as sorts and merges, may also be specified. The systems analyst or programmer may be assigned this coding job.

The programmer's responsibility for his program does not end with the final test. Changes, such as changes in tax rates and other constants or in procedures within the organization, may be necessary. The programmer is usually responsible for maintaining his program as long as he is with the organization. Since problems with a program may arise after a programmer leaves the organization, documentation is important so the program can be maintained by other personnel.

Documentation: When a program is completed, the programmer is responsible for producing a final documentation package about the particular program which has been coded. (See Chap. 10.) Program documentation is similar to the documentation of the systems specifications; however, it will contain sample output, program listings, logic flowcharts, and directions for operating the program. Documentation completes the programmer's role with a program.

SPECIFICATIONS TO THE PROGRAMMER

Before the programmer begins his job, the systems analyst will present him with certain specifications. The specifications are usually discussed at a meeting between the systems team and the programming section or with individual programmers who will code the program runs within the system.

First, the programmer should know where his program fits into the system. The output of other programs may be the input into his program. Similarly, the outputs of his program may become the inputs to other programs. The management of the computer programming section will usually supply the programmer with a standards manual which should specify standard methods for accomplishing certain programming jobs. Also, the standards manual will inform the programmer as to what routines have been prewritten, such as conversions of dates and other routines which can be used.

Usually, the programmer will be told what computer language to use. Most installations have a common standard that COBOL is the only language that will be used unless management approves other languages. Very rarely will the programmer be allowed to select any language he desires.

The programmer is also told what storage medium his input or output files will be upon and what methods of access and organization he will be using. Of course, the programmer interacts with systems analysts and designers and suggests or comments on how the system might be improved.

The programmer is also given detailed file layouts, input card layouts, and printer chart layouts when they are applicable. The programmer should have a copy of the systems documentations package, including the list of the data items within the system, their characteristics, what particular input edits can be used, and other procedures which insure accuracy and control. With this documentation, the programmer is given the standard systems codes and other standard names for files and programs within the system. The programmer is mainly concerned with the input-output specifications and the logic the program must carry out.

PROCEDURES WITHIN A SYSTEM

The procedures carried out within a system are generally the following:

1. Load data into files or computer memory
2. Retrieve records from these files
3. Process the records
4. Produce output records and other types of output

The programmer will be concerned with coding the most efficient methods of accomplishing these four steps. One method which he uses that the systems analyst and designer should become familiar with is the modular concept in programming.

THE MODULAR CONCEPT IN PROGRAMMING

The modular concept of programming essentially means subdividing the logic of the program into logical parts which are usually called modules. To effectively separate the program into logic modules, the programmer must understand the parts which make up the program. There are sets of instructions which the computer can carry out, functions and macro routines which the programmer can use, and independent modules which have been prewritten by other programmers within the organization. The use of these operations is usually through the concept of coding a main program which uses logical modules.

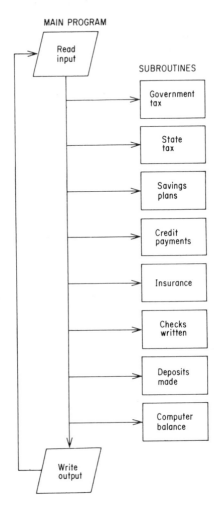

Fig. 8-3   Modular programming

For example, a banking system must produce account statements; it also might keep records and deduct amounts from checking accounts for federal and state government, savings plans, credit payments, insurance programs, and provide for bank accounting functions. Therefore, the bank program may be divided up as shown in Fig. 8-3. A mainline program shows the order in which the functions will be carried out. The functions themselves, such as deductions for credit payments, are placed in independent logical modules or subroutines within the system. As the mainline program is executed, the computer performs the subroutines.

When the modular system is used, it is much easier to maintain a program, to find errors in logic, and to make changes later in the program. Both analysts and programmers realize that the programs or computer runs within the system are not designed just to be implemented at the outset of the running of the system. At some time, programs will be changed, maintained, and certainly must be documented and tested. The modular system simplifies correcting and maintaining programs as well as testing them to see that the objective output is produced. It also provides better documentation than a nonmodular program.

DEVELOPMENT OF LOGIC

How a particular computer run should carry out its processing functions is one of the creative tasks of programming. There are many ways to carry out processing; and although different methods may both produce the same output, one may take twice as long as another. Some methods may be more accurate than others. Therefore, the logic that is developed should be the best and most efficient that can be used in that particular computer system. Probably the best approach is to develop an overall view before considering specific details. For example, in an inventory system, it would be appropriate to develop a general outline of the program logic before considering the actual details of stock deductions, reorder checks, purchasing requirements, and other such functions. Once the general logic is developed, then the programmer can develop the computer run in detail. Flowcharting has already been presented as a method of specifying program logic and flow. Computers are sequential machines in that they process in sequential order unless their path or sequence of processing is altered. Since computers do work in sequential order, sequential lists of steps can be developed for the system.

An algorithm is considered an unambiguous step-by-step procedure for performing some particular function. For example, the deriving of a square root from a number can be studied as a step-by-step procedure and could be called an algorithm. More complex algorithms such as those which find unknown roots of equations and accomplish other mathematical functions are very common in mathematics and management science. Lists of algorithms are available for many different types of functions. When they are not available, the programmer must develop his own step-by-step procedures for accomplishing the particular program function.

By developing algorithmic lists, the programmer can begin to see how the computer will carry out the functions.

As logic is developed, certain checks can be made to see that it is complete. These are as follows:

1. Initial value
2. Input-output functions
3. Procedures and processing
4. Logic
5. Loop control
6. Exit
7. Last records
8. Error routines
9. Continuous flow

Initial values: The programmer should consider what initial values should be programmed into the system. Initial values may be constants, such as tax rates, insurance rates, etc., or zero for total areas. Most computer systems do not automatically set any initial values at the beginning of the program run. Therefore, the programmer must be sure that areas within his program are cleared to zero or blanks if appropriate.

Input-output functions: The programmer should consider input-output processing: the points in the program where input records are brought into the system from peripheral devices, such as magnetic tape, disk or card, and where output is produced.

Procedures and processing: Once input-output is determined, processing procedures, such as calculations and the movement and transfer of data, can be developed.

Logic: Logic and altering program paths are important programming considerations. For example, a code value used in a program may determine different steps of logic, Therefore, the programmer should consider all the various sequences the program may take. This type of logic changes or alters the sequential path of the program. Whether a path is altered within the program during a particular operation usually depends upon an input value.

Loop control: Loop control or repeating of steps within the program is the real power of computer processing. In other words, if a hundred employees' payroll checks are to be processed, basically the same operations will be carried out for each of the hundred employees. The repetition of the steps which the computer carries out is called a loop. From the first input record to the last output record, there will be some loop within the program which will repeat these steps for each record.

Exit: An exit from the loop also must be considered. In other words, the loop must be terminated. If it is not, an "endless loop" occurs, which will continue to process until the computer is physically turned off or until the system runs out of input data. At that time, it will simply halt. Usually, an exit from a processing loop is made by some checking of an input value or by determining that the last input record has entered the system.

Last records: The programmer also must consider what will happen in the program when the last record is encountered. The last input record of a system does not necessarily mean that the programmer has finished with the program. After the last record is entered, there may be totals to accumulate and print out, or messages such as record counts and other such routines which are performed only after the last input record has entered the system.

Error routines: When the programmer specifies processing, he should consider error conditions that will arise in the system and what to do about them. The handling of errors is usually coded in an error routine. An error routine can be the printing out of a message to the system operator that an error has occurred or it can be some attempt to correct the error.

Continuous flow: When handling errors, the programmer should be aware that continuous computer processing is desirable. Any time the computer system has to stop because of an error condition, computer time is lost which means time and money are lost. Therefore, during all types of processing, the programmer desires continuous operation, and error conditions should not halt the system but should terminate the program so that the error can be checked while other programs are running. The program can be re-entered into the system later.

TYPE OF PROCESSING

The systems analyst will specify the types of processing needed by the programmer, such as when master files must be updated, when summarization of data must occur, etc. All of these types of logic will become the program steps coded by the programmer. The methods of updating files, that is, matching the transactions against a master, is a basic processing method in all systems. Usually, this is called file updating. Both the analyst and the programmer must understand how the files are matched and what logic is used to update a file. A simple flowchart for updating a file is shown in Fig. 8-4.

In addition to updating files, the analyst must know when data should be extracted from files to improve system efficiency. For example, if a file contains 50,000 records and only 600 of these records are needed to produce a certain type of output report, the analyst must know at what point these records should be extracted or when the entire file should be used to produce this particular report. Usually, records should

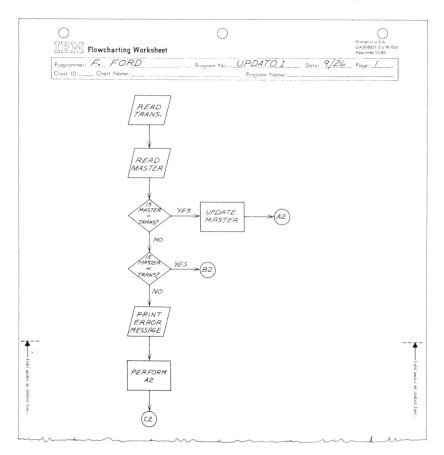

Fig. 8-4   Flowchart for update program

be extracted at the same time some other type of processing is being done. Therefore, a separate run is not needed on all 50,000 records of the file to produce reports which require only 600 records.

Using summarization methods can improve system efficiency. All the data that has entered the system for one month need not be processed simply to produce a report that summarizes the month. Summary totals can be calculated daily or weekly; when the report is needed at the end of the month, the summaries of the weeks can be used instead of all the data which has entered the system for the month.

In many systems, statistics about data conditions within the organization are needed. The systems analyst and programmer should have a good understanding of the statistical routines that can be used, such as standard deviations, correlations, etc. Basic statistical packages are usually furnished by the manufacturer of the computer and may be used by the programmers when the system requires them.

SPECIALIZED TYPES OF PROCEDURES

Several specialized procedures may be used within the system. One of these, table handling, is the processing or accessing of values within a table, usually called an array or a matrix. In some cases, the computer language used will be able to handle the table without sequential searches of the table. In cases where searches through a table are needed, a binary search may be used. In a binary search, the table is simply divided into halves, then quarters, and then eighths until the desired figure is reached. Tax rates, insurance rates, and other such constants are many times placed in tables in the computer's memory. If a table is used, one of the programmer's first jobs is to turn to the reference manuals for the computer system to gain a better understanding of how the system handles tables or how tables may be handled by the programming language.

Prewritten routines integrated into the operating system will usually be supplied by the manufacturer. They should be used when feasible as they save programming time and usually have been thoroughly tested and are very dependable. Procedures such as statistical analysis, probability functions, specialized mathematical routines (calculation of cosine, sine, tangent), and other such procedures usually will already be prewritten and available within the system for the programmer. Prewritten routines may come in computer-readable form or they may be simple listings in technical books. Also, many libraries contain books of programs that have already been written to accomplish certain common jobs. Usually, these routines are called functions or subroutines.

ERROR CHECKS AND HALTS

As stated before, the system should not be halted for errors, as the most desirable method of processing in a computer system is continuous processing. Most computer systems produce error messages which will be displayed to the computer operator as they occur. In other cases, the programmer must code error messages. Some techniques that the programmer can use to prevent errors or improve accuracy are:

1. Rounding of numbers
2. Overflow conditions
3. Significant digits needed
4. Negative and positive checks

5. Zero balances
6. Limits of accuracy
7. Translation accuracy
8. Cross-footing
9. Proof figures

Rounding of numbers: Most computer systems do not automatically round off calculated values. Therefore, the programmer must pay special attention to include computer instructions which will round values. A simple method of rounding cents is to add .005 to the value. COBOL has a special word called "rounded" which must be used to obtain rounded values.

Overflow conditions: Overflow occurs where not enough locations are left for total values. For example, if three memory locations contain the value of 999 and one is added, the answer will be 000 because the four digits of 1000 will not fit into the original three-memory locations designated by the programmer. This is called an overflow or "on size" error condition. It can be avoided if the programmer formats enough memory locations for all possible answers. Sometimes the possible answer is not known; therefore, the programmer must make checks for overflow using instructions available in the computer language. When overflow conditions occur, the value is always in error and is of no use in the system.

Number of significant digits needed: Computer systems are limited as to the number of significant digits that can be obtained. Although the actual number produced on an output report may be 10 digits long, perhaps only the first seven or eight digits will be accurate. Some systems have more significant digits than probably will ever be needed; however, the programmer will want to be familiar with the limits and number of significant digits of the particular computer system for which he is designing the application. Additional significant digits may be obtained by special programming procedures. Sometimes special control cards can be entered into the system to expand the number of significant digits which will be produced.

Negative and positive checks: During processing, the programmer can include negative and positive checks of values. In some computer languages negative values of numbers will not be kept unless the programmer makes special designations in his program. Many values are always negative or always positive; therefore, negative or positive checks of these values can be a check as to the accuracy of the data.

Zero balance: In many accounting systems, two totals or other numbers should always be equal. Therefore, subtracting one number from the other should always give an answer of zero, which is called a zero balance. In all accounting systems and other types of processing where it is applicable, the programmer should make this zero balance check and print out a zero balance at the end of his report to assure that all the values that should balance have balances within the system.

Limits of accuracy: Computer systems usually have a numerical limit to numbers that can be represented. Some computer systems have different limits on integers and reals. For example, if the limit is 10^{48}, any number exceeding this limit will be inaccurate. The programmer will want to know the limits of accuracy of the particular computer system he is using and analyze whether any values within the system may exceed this particular value.

Translation accuracy: Every high school student knows that $1/3 + 1/3 + 1/3 = 1$. However, when the fraction is transferred into decimal form it is easily seen that .333 + .333 + .333 = .999 and not 1. This type of error is called translation error and occurs in many computer systems. For example, a computer system may represent the decimal .1 (1/ 10) as a binary number. Therefore, .1 + .1 + .1 + .1 may not be equal to .4 but presented by the computer as .399999 because of the translation of .1 to binary. The amount of error that will occur is a mathematical problem which is beyond the scope of this book. However, in most cases, translation error can be overcome by presenting numbers to one more decimal point accuracy than is actually desired or by adding + .005 to achieve two-decimal-place accuracy, + .0005 for three-decimal place accuracy, etc.

Cross-footing techniques: Using cross-foots can assure accuracy within a system and check upon the inclusion of all calculations within the programmer's codes and steps. For example, three columns of figures A, B, and C should equal horizontally and vertically to the same values. Adding all of total A, all of total B, and all of total C should be the same as the total of A + B + C across the page. A simple cross-foot check appeared in Chap. 6. Although the values may not be printed out as output, they can be checked by cross-foots internally within the computer system by simply adding the values in the program.

Proof figures: Proof figures are values which can be added to the system to control limits and parameters within the system. For example, if no item in inventory is known to exceed $5.95, then 5.95 can be the proof figure. Therefore, the multiplication of any quantity by its unit cost should not exceed the multiplication of the same quantity by 5.95. Also, the difference between these two products should equal the total of the first figure subtracted from the total of the proof figure.

The more checks the programmer can make upon internal processing, the more accurate the output will be. All the calculation techniques listed above should be used when appropriate.

LOGIC FOR OUTPUT REPORTS

Producing output reports requires some specialized logic in processing the program. For example, in producing some reports the data must have some order. An alphabetical listing can be obtained efficiently only if the sequential input file is in alphabetical order. Sequential processing and sequential order within the file is a basic method of processing computer data.

Output reports that require summaries of certain data items, such as branch summaries, salesmen's totals, warehouse totals, regional totals, and the like, necessitate a control break. A control break is the change of any sequential identification number; for instance, if the program is processing warehouse totals, and the warehouse number changes from a 1 to a 2, totals for warehouse 1 are printed out and the program begins processing for warehouse 2. When the sequence changes from warehouse 2 to warehouse 3, totals for warehouse 2 are printed, and the program begins processing for warehouse 3, etc. The programmer must understand the process of control breaks, although he does not specify them. This is the system analyst's job. The analyst must also be sure that the data file is sorted into the proper sequence so that control breaks can be programmed.

MANUAL TASKS

The systems analyst is also responsible for procedures which include manual tasks. Usually, a systems flowchart contains specifications detailing what manual tasks must be carried out, and a job description is included for personnel. Specifications of the tasks to be performed can be listed in the job description.

Manual tasks may include hand carrying and moving data from one part of the organization to another, the distribution of output reports, clerical tasks of typing, photocopying. Tasks essential to system operation should be diagrammed on the flowchart. Simple tasks such as separation of carbon paper from the report, binding, and mailing should also be in the systems flowchart. The efficiency of such manual tasks is usually the responsibility of an office manager rather than the systems analyst.

CONSIDERATIONS FOR PROCEDURES AND LOGIC IN A REAL-TIME SYSTEM

Procedures and logic in real-time systems must have a program or some other control routine to handle input and output. This routine is both specialized and complex and is usually provided by the computer manufacturer. The routine handles data as it is received from communications lines. It details how telecommunications line and other system codes are used, and is also critical as to time factors, that is, the receiving of data from the telecommunications line without losing any data within a certain time period.

Input and output of real-time systems is usually qued, that is, data is ordered or given some priority as it is received or sent. Other than these considerations, procedures for real-time data are basically the same as those for other types of computer data. In other words, all the input edit/output edit techniques and procedure calculation techniques are used in real-time systems. When processing real-time data, common computer languages such as COBOL and FORTRAN require some intermediate step to be integrated into the real-time supervisory routine. Therefore, the assembly language provided by the manufacturer is ordinarily used to handle real-time information. Using macros provided by the manufacturer helps the programmer in programming for data which comes from communications lines.

SUMMARY

Procedures and logic refer to the programming necessary to make the computer system function as desired and to the manual tasks needed to operate the system. Although the systems analyst must participate in the programming phase, programming is usually the responsibility of a programmer. A program consists of step-by-step instructions which the computer system will carry out.

Basically, the programmer's job contains six steps. These are problem analysis, flowcharting, coding, testing, implementation, and documentation of the system.

Programming the computer runs is a separate phase from the analysis and design of the system. However, the specifications written in the analysis and design phase are passed on to the programmers so that they will have all the information needed to accomplish the programming of the system. Information as to where the program fits into the system, a rough idea of the logic and formulas that will be used, what standards must be adhered to, what programming languages will be used, what storage media will be used, the detailed file layout, input layouts, printer charts, and all systems data item codes and standards are presented to the programmer by the systems team. Essentially, the programmer should have a complete systems documentation package to refer to as he programs.

Basically, the programmer carries out four functions: loading the data into the files, receiving records from the file, processing these records, and producing output reports or other types of output. The programmer can best accomplish this job by using a modular concept. The mainline program states the general functions of the logic, and logical modules or parts of the program carry out the detailed functions of the program. The systems analyst will be concerned with logic of processing, of updating files, extracting from files, summarizing files, and producing statistical analysis. His ideas in these areas are passed on to the programmer.

The programmer is concerned with specialized procedures which he must accomplish, such as table handling, error checks, error routines and messages, whereas the systems analyst is concerned with how the program run fits into the system and the manual tasks which must be carried out to make the system function.

Preparing output reports requires sorting and the ability to match records and to recognize control breaks.

Procedures which must take place in a real-time computer system are essentially the same as the procedures in any other program once the input and output function has been accomplished.

Systems development is both an art and a science. The creating of logic which will process data within a system is probably an art, mainly because there are many different ways of accomplishing tasks, and the programmer will use the most logical and efficient method for producing valid computer runs. Procedures that have been proven through experience should be used. As the programmer is accomplishing his task of stating the detailed steps to the computer system, the systems analyst can analyze the system as it has been designed to assure that control and feedback will be accomplished within the system.

Questions

1. How does the systems analyst relate to the job of computer programming?
2. Discuss the programming cycle.
3. What must the systems analyst supply to the programmer?
4. Why should the modular system be used in programming?
5. What is meant by loop control?
6. What are error routines?
7. Discuss the rounding of numbers, overflow conditions, and number of significant digits needed as accuracy checks.
8. Explain the control break.
9. What are the basic four functions of a computer program?

Problems

1. Obtain a computer program from the library or the computer center and check its accuracy by applying the nine checks of logic and the nine error checks and halts to the program.
2. Lay out a report of sales by salesmen in various states. First, sort the report by salesmen within states. Then lay out another report sorted by states within salesmen. Which report reads the best? Where are the control breaks?
3. Draw a program logic flowchart that depicts the checking of inventory levels below 1000 units. If the inventory amount is below 1000 units print out the inventory number. If the inventory amount is equal to or above 1000 units print out the words "O.K." Input is inventory number and inventory amount. Output is the inventory number or the words "O.K."

9

Control

Control is essentially the restraints upon or the changes to a system which direct it toward its objective. In a computer system, the engineering design makes control to some extent automatic. In systems design, the systems analyst must plan control, and the systems users and managers must accomplish it.

This chapter analyzes control procedures and specifies control measures which the systems analyst can design into a system.

GENERAL STEPS IN CONTROL

Unless they are chaotic, all systems are controlled in some manner. The steps that establish control are:

1. Set an objective
2. Monitor the results
3. Compare the results to the objective
4. Adjust the system toward the objective
5. Repeat steps 2 through 4 continuously

These control steps can apply to all objects which have some function. For example, a thermostat is a very simple control device. The objective is the temperature desired in the room. The thermostat monitors the actual temperature in the room, and the result is compared with the objective which is set on the temperature gauge. The system then turns on either heat or cold air to adjust the system toward the objective. This process must be repeated continuously if the system is to be controlled continuously.

This principle of control is common to many appliances. For example, a television set is controlled by a channel selector, dials, volume controls, etc. Stoves and dishwashers are controlled by dials set by their users. Automobile controls include brakes, accelerator, shift, etc., operated by a driver. In all cases, the objective of control is to maintain the functioning or output of the system to satisfy the requirements of the user.

Control directly depends upon feedback mechanisms. Consequently, it is necessary to monitor the results or functioning of the system. Such system monitoring constitutes feedback, where the system's actual functioning is compared to its objective. In an automobile, for example, feedback devices would include the gas, oil, and temperature gauges, which tell the operator if the vehicle is not functioning

151

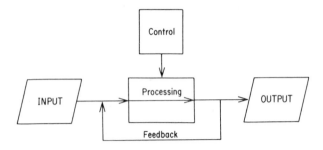

Fig. 9-1   The systems model

normally so that he may take some action to adjust the automobile system. Feedback normally monitors output. Figure 9-1 shows the systems model presented in Chap. 1. Feedback comes from monitoring output produced by the system before the output leaves the system. It can be fed back through the system or corrected at that point. In many cases, the feeding back of output means to return it to the input point for reprocessing. For example, in an assembly line any items which do not meet the quality control objectives are returned to the assembly line for re-assembly.

In a computer system, the amount of control to be applied can vary depending upon the importance and volume of the output. Although the application of control is usually expensive, critical items should have as many controls applied to them as possible. Previous chapters presented the techniques of controlling accuracy, input, and output. Some of these same techniques can also be applied to the system as a whole. This will be discussed in detail below.

BASIC TYPES OF CONTROL

The basic type of control applied to organizations that use computer systems are external control and internal control. External control refers to the laws, regulations, and specifications which exist as part of the environment. For example, state and federal laws are external controls, and the systems analyst can do little about them other than to adjust to them as efficiently as possible. Frequently, government laws and regulations are part of the system, which may cause less efficient operation. For example, reporting unemployment pay and deductions to the state unemployment office is a part of the system that may involve an extra program run as well as the storage of extra data within the system. However, this process must be carried out to satisfy the law. The systems designer cannot eliminate this process within the system; however, he can design it so that the process is carried out as efficiently as possible.

Internal controls are basically plans, standard operating procedures, guidelines, and rules under which the organization and system must function. Most guides, rules, and statements of objectives are produced in general form by management. They are carried out and implemented by department heads. For example, a general guide within an organization might be that the computer system should be compatible with that of a branch location. This is a standardization of computer processing, but at the same time, it is also a restraint upon the systems analyst who may wish to use equipment which may not be compatible. Although the analyst usually must accept internal controls, they are open to discussion and debate by the systems team, whether major or minor items. Besides the internal control generated by top management and middle management, the systems analyst also sets control procedures to the particular system he is developing.

WHAT CONTROL SHOULD ACCOMPLISH

System control should provide several features which are desirable in the system and needed for it to function. They are: (1) reliability, (2) security, (3) accuracy, (4) efficiency, (5) auditing procedures, and (6) adherence to the company's policies.

Reliability: Reliability refers to the ability of the system to accomplish its objectives continually. This means that the system should not break down frequently because of faulty equipment or procedures and that deadlines and other system specifications will be met.

Security: The system should allow for security within the organization. Security can be achieved by applying appropriate control procedures. For example, passwords used within computer programs can prevent inadvertent or fraudulent use of data files. Manual control policies can provide physical security.

Accuracy: The system should provide accurate information and reports. The input and output techniques presented earlier help insure such accuracy. This chapter will present in detail some additional controls which can be used to promote a system's accuracy.

Efficiency: Efficiency is essentially performing the tasks in the best manner and producing the highest quality with the least effort or cost. Assuring efficiency within a system usually means that the systems analyst must consider in detail all the alternatives to achieving the system objectives. Efficiency can be promoted and encouraged by using appropriate controls. For example, deadlines and similar regulations can improve system efficiency.

Auditing procedures: System control should include an audit trail within the system. The audit trail enables the tracing of a data item from its creation to its final use. Specialized printouts and other system output are ways of providing an audit trail.

Promoting adherence to organization policies: No control applied to the system should conflict with organizational policy, and those that do apply should promote recognition of other policies within the system. For example, certain controls applied to producing accurate payroll checks will also promote the organizational policy of paying all employees on time.

MONITORING AND FEEDBACK

Control of any object depends upon adequate feedback. The systems analyst and managers cannot adequately control an operating computer system without feedback as to how that system is functioning. In many cases, adequate feedback is not provided for computer system operation. The system monitoring and feedback that can be produced are as follows:

1. Log books, log sheets, and control tickets
2. Systems labels and codes
3. Registers and printouts
4. Computer consoles and computer operators
5. User and customer complaints

6. Audits and consultants
7. Error analysis
8. Statistical sampling
9. Written specifications
10. Hardware monitors and software evaluators
11. Industry averages
12. Cost reports
13. Miscellaneous

Log books, log sheets, and control tickets: A basic method the systems analyst can use to provide feedback on a system's operation is logging at many points throughout the system. For example, log books used to enter the date data is received from user departments are controls over data movement within the organization. Assigning control tickets to batches of data as it moves through the system provides a means for identifying the source and use of data as it is processed. The computer operator log sheets show times required for particular computer runs and also indicate when data entered the system as well as when the last output was processed and distributed. Distribution log books show when the output left the system for its designation. All logging entries should be used by the analyst as monitors and feedback mechanisms to show system managers and management how the system is functioning.

Systems labels and codes: Systems labels, such as data sets, file labels, and other codes, provide feedback during processing and help control security and movement of data. For example, computer operators can visually check labels on magnetic tape or disks to determine if the proper tape or disk has been mounted for a particular computer run.

Registers and printouts: Periodic printouts of data files can be used for backup data and also provide visual proof and analysis of data contained on computer storage media. Printing out the contents of a magnetic tape containing inventory items is a way to sample whether the items are accurate and proper. Sometimes memory dumps (a printout of the contents of a main memory of a computer) will indicate the status of the system. The systems analyst should also provide for computer runs which produce certain types of registers, that is, a detailed printout about a certain area within the organization. For example, a journal register shows every entry made to the accounting system. A payroll register shows the weekly or monthly pay of every employee. Inventory item registers show the status of every item in inventory. Using registers and circulating them within the organization provides an immediate reference for items and eliminates formal requests for them from the computer system.

Computer consoles and computer operators: Computer consoles provide immediate feedback on the system's status. On most computer systems, light indicators display information as to whether the system is halted, processing, etc. Also, error conditions and other signals may be printed on the computer console. The console typewriter, which logs messages to and from the system, can also yield information about the processing of a system or a specific computer run.

User and customer complaints: User and customer complaints give the systems analyst and managers one of the best feedbacks. Naturally, customer complaints receive more attention than other complaints, as they may indicate something basically wrong with a computer system. Wrong billings, slow delivery, and other complaints should be analyzed thoroughly and the error remedied.

Persons within the organization who use output reports can also provide feedback to the systems analyst on the system's performance. Written specifications distributed to all systems users telling how to suggest system improvements should be an integral part of the systems specifications. User feedback can give the systems design team useful information to integrate into future systems or to correct the present one.

Audits and consultants: Periodic audits of systems provided either by Certified Public Accountants or internal auditing teams also give systems analyst information on the system's performance and on adjustments that might be needed to correct the system. The systems analyst should read audits very carefully and adjust the system to meet any problems specified in the audit.

Outside consultants are hired by many organizations to obtain feedback on systems operations. Outside consultants can sometimes perceive problems which those within the organization or on the systems design team overlook because of their closeness to the problem.

Error analysis: The best system is error free. However, methods should be set up to analyze errors if they occur within the system. There are adequate mathematical techniques for analyzing errors and determining the extent of error in operations. Formal mathematical error analysis may not be required by the organization; however, some method should exist to determine the percentage of error which occurs within the system. This particular percentage number can be used as the basis for future improvement of the system.

Statistical sampling: Statistical sampling techniques can help detect errors and also determine system characteristics. Random samples can be used to obtain information about average values produced by the system, such as the average amount of computer time used for runs. Sample techniques have been used on the assembly line for many years to provide quality control of products.

Written specifications: Almost all monitoring and feedback processes depend upon written specifications, which show clerical workers and other system personnel how to obtain and send feedback information to appropriate persons. Many times errors denoted in the system are never reported. Perhaps some manual task corrects the error before it leaves the computer center or after it reaches the user. By reporting these situations and by having written specifications as to how to report them, they can be corrected. Written specifications of the system are the basis for all the efficiency in system operation.

Hardware monitors and software evaluators: Several instruments now available can monitor the efficiency and performance of a computer system. In large installations, the price of a monitoring device is frequently paid for by improvement of the system. Similarly, software applications can be evaluated through check-lists or, in some cases, other software programs.

Industry averages: Comparing a system's performance with industry averages is a basic feedback method for assessing system performance. For example, if an organization in the five million dollar a year revenue class is spending sixty thousand dollars a month on computer operations, whereas its competition is spending only ten thousand dollars, the system is not being used in the most efficient manner. Although this example may seem exaggerated, many such cases exist. Comparing the organization's performance with average performances within the industry or service area of the organization can tell management if the system is efficient.

Cost reports: Cost accounting is a basic technique used by most organizations to control expenses. It is equally applicable to computer system operation.

Miscellaneous: According to the company's situation and its application and function, the system will contain many natural feedbacks. The systems analyst determines what these feedbacks are and how to capture and use the information they provide to adjust the system. Basically, anything which indicates the status or function of the system can be used as feedback for control. For example, not listed above but also important are reports on maintenance, messages generated by the system, and other feedbacks which indicate how the system is operating.

INPUT DATA CONTROL AND ERROR CONTROL

All the techniques studied in the previous chapter on input apply for controlling accuracy of input data. As stated before, input is a major problem in most computerized information systems. Some of the techniques listed in Chap. 7 are also listed here; however, these techniques are generally procedures rather than specific input editing steps. They are:

1. Sequential prenumbering of source documents
2. Reporting received data
3. Keypunching and verification
4. Batching of data items
5. Due dates for reporting of information
6. Item counts and control totals
7. User codes and key numbers
8. Listing of input and returning to user
9. Use of prepunched data cards
10. Documentation
11. Use of punched card forms as source documents
12. Control of input data after use
13. Editing techniques

Sequential prenumbering of source documents: In most cases, source documents used within the system should be sequentially prenumbered. This allows for both item control and batch control. Documents which are not preprinted can be numbered with a rubber stamp which rotates numbers sequentially. Numbering source documents gives a control of the documents, provides a check for the presence of the document, and allows for sequencing of all documents. In many cases, such documents as purchase orders, shipping orders, etc., may be numbered sequentially as they are issued. This numbering of the source document can become a principal or secondary key for identifying information on that particular document. It also allows for audit trails for both internal and external audits.

Recording of data that is received: Recording or logging in the date, volume, and other information about data received from outside the system keeps basic information which allows for both audit trails and scheduling of input preparation. At the time the data is received, personnel in the data preparation section should make a physical review of the data. Data which is out of order or incorrect can be returned for correction. This is essentially the feedback of errors within the system for reprocessing. However, in this case feedback takes place before the data enters the actual computer system.

Keypunching and verification: Hiring experienced keypunching or keyboarding operators will increase the accuracy of data as it enters the system. Most managers within the computer industry agree that it is better to pay higher salaries for better qualified personnel than to attempt to save money by hiring less experienced personnel, since the quality of the output or objectives of the system depend directly upon the accuracy of the input. Verification of input data assures a high degree of data accuracy, although it also doubles the work and the cost. Verification of data based upon the task performance of the key operator or by statistical methods can sometimes be used with good results.

Batching of data: Batching data as it is received and applying a batch control ticket to it helps both to control the data as it moves through the system and to pinpoint errors. Identifying data by batch number creates a control of the data as it is received at certain points within the system. Although batch data-processing systems are common, it is probably not the system of the future. Data collection is becoming increasingly random because remote data enters the system as it is created. However, until the collection of data where it originates is further developed, batch control is probably the best method of assuring adequate control over the data as it enters and leaves the system.

Due dates for reporting information: Assigning due dates, when data must be received for processing, is a practical control for maintaining schedules and meetings deadlines. Due dates are used in cases where hours worked must be reported to the data preparation section before payroll checks can be prepared. For example: "Only those time cards turned in by Wednesday afternoon at 5:00 P.M. will be processed for checks to be issued on Friday afternoon."

Item counts and control totals: Item counts and control totals have already been mentioned as a basic input control method and can be reviewed in Chap. 5. As stated before, control totals and record counts placed upon the batch control ticket can be checked before the data is distributed. If the counts do not match, then there is a probable error or loss of data within the system.

User codes and key numbers: User codes, key numbers, and passwords help control fraudulent or accidental use of data. For example, only data coded with a 329 is to be used in the inventory control reporting system. The number 329 is generally known only to the persons working in inventory. Therefore, data which does not contain the 329 code will not inadvertently be mixed up with other types of accounting information such as sales, statistics, or payroll data. (Codes are explained in more detail later in this chapter.)

Passwords are words known only to the users of certain data files. These passwords are changed periodically so that they remain confidential. As the computer system is accessing the data, only those programs having the proper password will be allowed access to the input.

Listing of all input and return to user: To control errors in data, many organizations place the responsibility for data accuracy on the system user. Therefore, the data preparation center simply prepares the data, a computer listing of the data is made, and the listing is returned to the users. The users are responsible for checking the accuracy either in detail or by some sample technique. This method also assumes that the user is responsible for the accuracy of output, although processing techniques used by the programmers will affect the accuracy of the output.

Prepunched data cards: Prepunching data cards through the card-punch device attached to the computer system eliminates the preparation of large amounts of data which must re-enter the system. For example, most oil company credit card billings are sent on standard punched cards in which the basic information has been punched by the computer, such as account number, date, etc. Only the amount paid when the consumer pays his bill has to be punched in the card before the data can be entered into the system.

Documentation: Documentation is the basis for controlling all aspects of the system, especially input. The systems analyst should prepare written instructions as to how input will be handled, how errors will be corrected, and so on. Careful attention must be given to the written instructions, since all handling of data as it enters the system is based upon procedures written in this documentation portion.

Use of punched card forms as a basic document if possible: Wherever possible, the standard punch card should be the form used as a basic document. Punched cards are easy to handle, and in many cases low volumes of data can be punched into the same card upon which the information is written and created. In the future, optical character recognition and direct terminals will probably replace punch cards.

Security of input data after use: Other than for audit trails and backup or historical purposes, all input data should be destroyed after a certain period of time to prevent its accidental re-use within the system. Since input data created in June for a billing cycle will look very similar to input data created in December, there must be adequate control so the older, already used input data is not re-used. Using different colored cards helps to control this particular problem. Also, in specifying time periods for retaining data, the systems designer should use industry standards or government regulations as a basis. Sometimes, after it has been used, input data is sent to the department that created it and that department is responsible for storing the data.

Editing: All the editing techniques listed in Chap. 7 should be applied when applicable to detect possible errors which may occur during data preparation.

PROCESSING CONTROLS

After the data leaves the input preparation section, it is moved to the computer center for processing. There are several methods for controlling processing. Most of them are general techniques and should be used in addition to the techniques outlined in Chap. 8. The general techniques include:

1. The logging and scheduling of processing runs
2. Restart and check points
3. Beginning and ending of run messages
4. Periodic printouts of files
5. Libraries for tape and disk
6. Labeling, external and internal
7. Table comparisons

Logging and scheduling of processing runs: All production runs made by the computer should be logged by either the computer operator or the control clerk. The logging of the data allows for controlling of the status of the data and also for information as to where the data is at any particular time. A simple log contains little more than the date received or sent and the name of the data which has been handled. A good computer-room procedure is to give input from the input preparation section to the control clerk, who logs the material in a log book. He then schedules it at some time when the computer schedule is open, keeping in mind the deadlines which this particular computer run must meet. The input data is then prepared for the computer operator, who picks it up at the control desk. It may or may not be logged in and out from the control desk. After the run is finished, the computer operator returns the input data, output, and any output records to the control desk. The data is again logged, as processed, and is given to the distribution section, which prepares it for distribution throughout the organization.

At the computer console, the computer operator may also keep a log book to document the use of the computer. In other words, for all 24 hours of the day, the log will show whether the computer was running or idle, the running time of a particular program, or whatever information is desired about the computer's use. The log also allows the checking of processing runs and is a good feedback mechanism for computer center managers to analyze computer scheduling.

Restart and check points: Processing interruptions and other error conditions or machine malfunctions during processing runs can be controlled by using "check points." The check point is nothing more than a recording of the status of the computer at certain intervals. Without a valid check point, an entire run might have to be repeated because of an equipment malfunction. By recording the status indicators of the computer—such as the memory locations, the instruction address in the program, what records the tape or disk is using—the status of the computer can be obtained, for example, every 30 minutes. Thus, the most time that can be lost is 30 minutes, for the computer operator can simply restart the run at the last check point. This is usually not an automatic process, however, but must be planned for by the systems designer or the programming section.

Beginning and ending of run messages: A good computer center procedure is to have all programs print upon the console typewriter (if available) a message which denotes the beginning and ending of the program run. Logging these messages increases efficiency, for the computer operator is signalled at the end of the run and can be sure that the next program is ready. One of the objectives of an efficient computer system is for continuous processing, rather than for numerous halts and starts. On very long runs, it is also a good idea to print out messages indicating the progress of the program.

Libraries for tape and disk: No matter how small the organization, it should establish a library for data files. In many cases, it is also desirable to have libraries for programs, utilities, and other routines used for computer processing. Libraries should be controlled by one person. A tape or disk library can be run very much like an ordinary library. Data tapes or disk packs are checked out and the borrower is responsible for them until their return to the library. In most organizations, the magnetic tape or disk library is a fireproof file or vault.

Labeling, external and internal: Clear, concise, and complete external labels enhance the efficiency of running a system. All tapes, reels, and disk packs should be labeled. The label should include the name of the data file, the date it was created, the date it is no longer useful and, in many cases, the programs that use the file and those that create it. The department which uses the tape and disk should be designated to prevent misuse of files. Volumes of data are generally identified by gummed labels. Without an external label, it is impossible for the computer operator to tell what is on a tape without first dumping it onto the printer. (Internal labeling is discussed later in this chapter.)

Table comparisons: A computer operator is frequently asked to enter data into the system. Usually, this is low-volume data, such as the current date, the current tax rate, or some other constant for the program, which changes frequently. The validity of the operator's input must be checked by the program, and this is usually done through a comparison with tables internally in the program. For example, if the operator enters the date, the month should be checked against 12 possible numbers or combinations, the year against the current year or the next five current years, and the day against a value of 1 to 31. In other words, computer operator input should be edited for errors just as regular input is edited.

All of the above processing controls should be part of the written systems specifications which the systems analyst and designer prepare. Controls can be added as the implementation of the system begins and after some operating experience with the system.

OUTPUT CONTROL

Once the data has been processed, there must be ways to assure that the output is correct and that it reaches its proper destination. Techniques that have been proven over the years through experience are:

1. Proper scheduling
2. Logging of all reports distributed
3. Separation of personnel handling output and input
4. Exception reporting to the computer operator
5. Sequence checks
6. Physical security of output
7. Backup copies—"grandfather, father, and son" concept
8. Control totals and control figures
9. Written directives for output distribution

Management may also add controls to output.

Scheduling: The computer system must be scheduled to meet deadlines for certain reports and to avoid peak periods, when the computer center is saturated with service or processing requests. Service essentially means providing a fair share to all of the users and systems applications which are running on that particular computer system. Although scheduling is basically a function of the computer center management team, the systems team will want to pay particular attention to scheduling of reports within its systems design. In other words, 10 or 15 reports should not be scheduled for processing on the same day unless there is adequate computer time. The first of the month is usually a critical time for computer processing because the accounting cycle demands many reports and output be produced at that time. Also, the systems designer will want to pay particular attention to the company policies, such as due dates for reports, government tax returns, etc. Before designating due dates for output reports, the designer should check the computer time schedule currently in use.

Logging of all reports distributed: Although rarely done in most small computer centers, it is a good management technique to log all output processed and distributed within or outside the organization. Log books serve as audit trails as well as a control feature which can be checked when reports are missing or late. Logging can be done by the computer operator, control desk personnel, or distribution personnel of the computer center. Normally, the log need only contain the date, the name of the report, and the number of copies distributed to keep a proper record of what the system has produced.

Separation of personnel handling output and input: For certain critical output, such as classified documents, payroll checks, vendors' payments, and the like, it is a good control technique to separate personnel who will be handling input from the operator who will be handling output. This is a well-known management technique to prevent fraud and provide security of data or information.

Exception reporting to the computer operator: A management technique which helps inform the manager of unusual conditions within the organization, exception reporting can also be used to inform the computer operator of unusual conditions in the system, especially at output times. Exceptions can be reported to the operator by programming techniques whenever unusual conditions occur. For example, most detail listings will read and print the same number of records. Therefore, output reports should contain the same number of printed lines as there are records in input. Any exception can be reported to the operator, usually through the console typewriter. The operator is not expected to correct errors, but he is prevented from distributing erroneous output until a systems analyst or programmer has checked the reason for the exception. The systems analyst usually specifies exception checks to the programmer when the systems specifications are turned over to the programming section. However, some programmers through experience will automatically add these controls without any special specifications from the analyst.

Sequence checks: Sequence checks can be used during input processing or output. To control output, a sequence check basically checks that all output has been produced. For example, if the organization prints 95 payroll checks a week, checking the beginning and the last printed check numbers can determine if the correct number of checks has been produced. Sequence checks often may be approximate values rather than exact; in some cases, the system can be designed so that control checks can be made by comparing separate runs of the same information. For example, the inventory item listing produced for a warehouse may be compared against inventory reorder listings prepared for the purchasing department. Certain figures such as purchases and received stock may be the same for both of these reports so any difference can be checked before the report is distributed.

Physical security of output: It is only common sense to provide adequate physical security for storage files and certain types of output. Most computer centers use safes and vaults to store magnetic tape and disk packs and filing cabinets, possibly fireproof, to store output reports, card decks, source documents, and other data. Although physical security is mainly a management function, the systems analyst may want to include certain equipment specifications for physical security in his systems design.

Backup copies—"grandfather, father, and son" concept: Output produced on some mass storage device such as magnetic tape or disk should have suitable backup copies. This is usually called the "grandfather-father-son" concept. When a magnetic tape is updated, the tape used as input becomes the father, and the new updated tape output is the son. As the next cycle is run, the father becomes the grandfather, the son tape becomes the father, and the new updated tape becomes the son. Usually, three generations of backup copies are all that is necessary. In most cases, these tapes or disks are kept in separate locations. Therefore, any loss or destruction of output can be recreated by re-running the father or grandfather file.

Control totals and control figures: A basic way to control output accuracy is to use control totals and control figures. Control totals are the totals which accompany the input data on the batch control ticket. These figures, such as total invoices or total sales, can be checked against the output billing statement totals or output sales report totals. Any discrepancy can be checked before the output is distributed. Control figures can be both approximate and exact. If the output distribution section knows the approximate totals for output reports, these totals can be checked before distribution. Any large differences in the figures would be reported as a possible error. Record count, number of pages in a report, number of exceptions, etc., can all be used as control figures depending on the particular system.

Written directives for distribution of output: All commonly occurring cyclic types of output should have proper written instructions about distributing that output. This information should specify the number of copies, the persons who will receive the output report, those to contact in case of problems, etc.

In summary, the computer center manager and the operation section supervisor are responsible for most output control techniques. The systems analyst is concerned with designing the system that produces the output and perhaps determining where the output should be distributed. However, any techniques which the analyst can contribute to output control are usually welcomed and should be written within the systems specifications.

There are several methods for control of mass storage files. These methods are divided into four main categories:

1. Library methods
2. Physical outside labels
3. Internal magnetic labels
4. Backup and physical security

The operating system takes care of the internal program library which resides on the tape or magnetic disk. In a large computer installation, the physical library is usually under a librarian's control. All the standard library techniques should be used, including checkouts and return privileges.

Physical outside labels should consist of the file name, the volume name, the reel number or how many reels, serial number, date created, date no longer valid, density, and the system or department where the data set is used.

Internal labels should consist of the file name, the reel number, the date created, the date no longer valid, and record count.

Backup and physical security should consist of grandfather-father-son techniques, duplicate copies, separation of storage facilities, and the like.

HARDWARE CONTROL

It is important that the equipment or hardware used in the system should operate properly and without error. Most of the equipment currently available has been engineered to be practically error free. Errors in computer hardware are so rare that they are not a major consideration in systems design or control. Hardware control techniques are built into most computer systems and usually are adequate to assure accuracy or reliability of the equipment. Equipment does sometimes break down and must be maintained; however, other than "down time" (the time the equipment is not operating because of some breakage or malfunctioning), there are few equipment problems. Errors found in computer processing are usually the fault of systems design and programming. In most cases, no errors can be contributed to the hardware.

The following are hardware control techniques:

1. Magnetic labels, counts, codes
2. Parity bits
3. File protect rings
4. Backup equipment
5. Read-write and write-read
6. Read-twice and write-twice
7. Periodic maintenance
8. Redundancy
9. Validity checks, such as valid op codes, valid addresses, valid characters
10. Echo checks where hardware sends completion signals
11. Use control, that is, checks to see if the device is being used
12. Hole counts
13. Cyclic checks or cycle checks
14. Longitudinal parity and diagonal parity
15. Boundary detection

CONTROL BY COMPUTER OPERATOR

The computer operator is essential for proper control during processing and running of the computer. The systems designer is responsible for specifying the particular messages and other techniques which can be used to signal an error to the computer operator. Some basic operator messages that should be specified in all systems are:

1. Ready messages and end-of-job messages
2. Control values
3. Messages that files do not match
4. Systems messages
5. Object time messages

All of these messages should be documented and available in the computer run book.

Ready messages and end-of-job messages help the operator determine the status of the system and what programs are being processed at any particular time. Because of the continuous flow of jobs and processing which take place in a modern computer center, it is difficult to determine what is going on internally in the computer at any time without some logging or messages to indicate the program's start and finish. In some systems the Job Control Language automatically takes care of this. (See page 000.) However, programmers should be directed to include a message at the beginning and at the end of the program, to signal the operator.

As any program ends, certain control values should be printed out as messages to the computer operator. The technique of using these control values has been discussed earlier. Therefore, documentations should be provided so that the operator knows the exact meaning of the control value printed and the method of checking its accuracy.

In most computer processing, master and transaction files are matched; that is, a master file is matched to the transaction which updates the file. The operator should always receive a message when files do not match correctly. The action the operator takes if this message is printed out is determined by the processing being done. However, an incorrect matching of files should not escape the operator's attention. Since it is an internal process, the operator can be aware of this only if the programmers code error messages in the program.

Systems messages and object time messages are messages to the operator which are programmed into the computer operating system by the computer manufacturer. The reference manual for these messages should be available to the operator. The systems analyst and programmers should include in the systems documentation important messages which might occur in the system. Examples of systems messages are the wrong label on mass storage data files, peripheral devices not working, etc. Object-time messages are errors which occur during program execution, usually because of an error in data.

SYSTEM CODES AS A CONTROL TECHNIQUE

Almost all systems use codes in their structure, such as names of files, program runs, procedures, etc. Proper structuring of codes enhances system control, because system components are identified easily. Similarly, lack of meaningful codes can cause lack of identification and thus, lack of control. System structures often coded are the organization the code applies to; the area in the organization; the department or branch; the type of application, such as accounting, finance, etc.; the system being used; the particular program run or processing step; and the type of data, whether input, output, or exception. Each organization and most likely each system will have its own coding method.

These methods depend upon the type of application and the type of organization using the computer. However, all codes should have proper structure and meaning. For instance, the code CR22 tells very little about the system. It could possibly mean computer run 22. But CR22 could be anywhere in the system and could be accomplishing anything. On the other hand, a structured code might read FCHIA3R. Although this appears difficult to read, the computer operator, programmer, or analyst can usually identify the system part this code applies to by breaking it down into elements. For example, in FCHIA3R the components are: F = finance, C = cost accounting section, H = Houston office, I = inventory application, A = analysis system, 3 = run 3 of that particular system, and R = report output. The code is not meant to be exact, but is rather an example of the types of structure that can be used in a system. Figure 9-2 shows a systems flowchart with naming of program runs and files within the flowchart by the use of structured codes. Notice the output file names and the computer run names. It is apparent that if FCHIA3R is an output report, it is produced by FCHIA3P, the program used to produce the report.

SYSTEMS CONTROL BY MANAGEMENT TECHNIQUES

The final responsibility for all organization controls, including the control of computer systems, belongs to top management. Many management control techniques have evolved over the years. The basic management functions which lead to good control are planning, organizing, communicating, and motivating.

Planning—forecasts and budgets—controls the financial part of the system and also determines the objectives necessary to systems development.

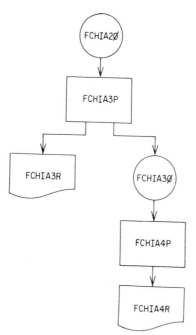

Fig. 9-2   Example of system codes

Organization structures have important impact upon systems development. Whether the computer manager reports directly to the president or whether he reports to an accounting manager can have great impact upon the ability of the systems analyst to interview, analyze, and move within the organization to determine systems design.

Providing adequate communications within the organization helps not only during the systems development phase but also during actual use of the system. Other important management techniques in systems development and computer use are the development of adequate standards and the keeping of proper documentation.

Motivation of personnel through incentives, career development, etc., will probably lead to better working methods and better control of the system.

Control, however, starts with well-defined objectives. According to the theory of control, control is the changing of a particular output so that it meets an objective. Unless these objectives are well defined, little control will exist within the organization.

SUMMARY

Control can be defined as any restraint or change in a system which directs it toward its objective. There are basic steps of control:

1. Set an objective
2. Monitor the results
3. Compare the results to the objective
4. Adjust the system toward the objective
5. Repeat steps 2 through 4 continuously

Control is applicable to almost all mechanisms. The objective of control is to maintain output of the system and satisfy the systems requirements or objectives.

The amount of control placed upon a system depends upon the importance of the data and output the system is producing as well as the volume.

There are many types of organizational control, external and internal. Basically control should provide reliability, security, accuracy, efficiency, and proper auditing procedures. Hopefully, control techniques also motivate employees to help the company meet its objectives.

All control depends upon the ability to monitor results and provide feedback so that the functioning of the system can be changed toward its objective. Therefore, certain monitors must be provided in a system. These are basically anything which indicates the status of the system or indicates how it is functioning. Systems analysts and designers are particularly concerned with data control, error control, and input, processing, and output control. Physical security and control of data storage is also a prime consideration. Most equipment and hardware controls have been engineered into the system and are a minor concern of the systems analyst. During processing, proper action by the computer operator provides the basic method of control. Therefore, the system should be designed to keep the computer operator aware of what occurs within the system, and the system should be documented well enough so that the operator knows the proper response.

The use of structured codes helps to monitor the system and identify elements within the system and provide feedback to persons using the system.

Finally, the responsibility for control rests with top management. By using good management techniques, the organization can be assured of proper control. However, lack of clear objectives of the organization will result in the lack of control over computer operating systems.

Control is not a static process; that is, objectives are not set for all time, but change according to the desires of management personnel. Therefore, when objectives change, control measurements and the system will also change to meet the objectives.

Questions

1. Name the five steps of system control.
2. Discuss the difference in external control and internal controls.
3. What does reliability mean?
4. What is feedback and how is it obtained?
5. Name six common types of feedback in a computer center.
6. What is a register as applied to a type of printout?
7. As applied to the five steps of control, what is a user complaint?
8. What is meant by the phrase "indicates the status of the system"?
9. Discuss three input data control techniques.
10. Explain an audit trail.
11. What is a password in a file?
12. Explain the use of prepunched data cards. How does this control errors?
13. What is a library as applied to the computer area?
14. What output control techniques should be applied to payroll checks?
15. Explain in detail the "grandfather, father, son" technique.
16. What should be on the external label of a magnetic tape file?

Problems

1. Write down the names of five common systems you are familiar with, such as a thermostat, a traffic light, etc. Explain how the first four steps of control are carried out for each system named.
2. Find a system within the organization you work for or the school you attend. Write a report on its reliability, security, accuracy, efficiency, method of audit, and how the system relates to organization policies.
3. You are the manager of the general ledger system, and the trial balance is your specific duty. Write a report to management about input data control.

10
Documentation

Managers, writers, and authorities in the computer field have been practically unanimous in agreement that good documentation is essential to the efficient operation of any system. They have also unanimously agreed that the quality of documentation in most systems is rather poor. It is difficult to determine why any procedure so highly recommended by so many knowledgeable people in the field has also been so ignored by many systems analysts, programmers, and managers. Perhaps the pressure upon computer personnel to produce valid output and to meet deadlines and budgets accounts for this lack. Or perhaps the lack of standards in documentation—accepted features or principles of what should be documented—have caused the managers to be lax in this area. The student of systems analysis, systems design, or programming, therefore, will want to focus on the systems phase known as Documentation. This will not only make his work easier and more efficient, but will also reflect in the quality of his work. Those who ignore documentation usually produce poor quality output.

Documentation is any writing, typing, printing, or other recording which is referred to for information about the development, operation, or use of a system or parts of a system. This chapter examines the various types of documentation and the various needs of personnel for documentation and explains each type of documentation.

TYPES OF DOCUMENTATION

Basically two types of documentation should be presented within an organization. *Standards documentation* is the documentation of standard operating procedures and standard codes, rules, and policies which should be used in all systems. *Operating documentation* is the documentation of each system the organization is currently using.

The manager who is responsible for systems analysis, systems design, equipment selection, programming, and computer operations should develop a standards manual. It should contain all the general rules, codes, and other directives which should be followed in developing all systems. Standardization has proved over many years to be important in almost all areas of development, from automotive engineering and electronics to construction and business management. Standards help the systems analysts develop methods which can be used for integrating and understanding systems. For example, although it occurs many times, it is ridiculous to develop one accounting system which uses the code 1, 2, 3 for marketing, sales, and finance and another system which uses the code m, s, f for marketing, sales, and finance. The proliferation of such differences in codes, structures, and methods creates problems throughout the organization and especially in the computer area. To emphasize further, programmers should know what languages they are allowed to use, what types of flowchart they should provide, keypunch conventions, methods of program testing, use of job control language, and many other standards which, if not specified, might differ in each program.

Systems managers and users must also have documentation for reference. Very fine systems have been implemented for computers, but are not used fully because there is no documentation to explain how to use the system. For example, an engineer may develop a very fine statistical processing system; however, the marketing, finance, and research people may never use it if proper documentation on its use is not available.

WHAT SHOULD BE DOCUMENTED?

Besides a general standards manual, every system which is being implemented on a computer should consist of six basic documentation papers. These can be briefly outlined as follows:

1. Management needs, such as objectives, narratives, plans, budgets, and time tables.
2. System and feasibility studies, the collection of documents from systems analysis, and feasibility phases.
3. Systems design, which is essentially the system and equipment specifications.
4. Programming documentation, which is the documentation of each program which runs on the computer.
5. The operational papers, usually called "run" books, which present the procedures for the computer operator to run the system.
6. The user's manual, which explains to nontechnical personnel how to use the system as it is implemented on the computer.

The rest of this chapter is devoted to these six documentations and the standards that should be set up before system documentation begins. The type of documentation used will vary from organization to organization and depends upon the type of application to be developed. Also, the computer system manufacturer furnishes substantial documentation, much of which can be relied upon when other documentation is not available.

STANDARDS DOCUMENTATION

A good standards manual should be developed before any systems analysis, design, programming, documentation, or use of the computer takes place. This manual may be called Standard Operating Procedures, or Standards for Information Processing, or Standards for Use of the Computer, or any other such title. Basically, it should follow this outline:

1. Introduction
2. Methods for requesting changes to standards

3. System designations
   (a)   Name
   (b)   Document number
   (c)   System steps
4. Plans for system studies
   (a)   Steps to follow to initiate a system study
   (b)   Approvals of system studies
   (c)   Assignment of personnel
   (d)   Checklist
5. Systems design standards
   (a)   Symbols
   (b)   Work sheets
   (c)   Codes
   (d)   Names
   (e)   Equipment use
   (f)   Formats
6. Programming standards
   (a)   Problem analysis
   (b)   Languages to be used
   (c)   Documentation needed
   (d)   Flowcharting
   (e)   Keypunch conventions
   (f)   Coding
   (g)   Testing methods
   (h)   Input-output
   (i)   Logic
   (j)   Use of sorts, merges, transfers
   (k)   Job control
7. Standards for documentation
8. Run book preparation standards
   (a)   Statement of program objectives
   (b)   Systems changes
   (c)   Operating instructions
   (d)   Paper used on printer
   (e)   Use of input-output files
   (f)   Labels
   (g)   Halts and checks
   (h)   Control totals and control figures
   (i)   Distribution of output
9. User's manual standards. Essentially the standards for developing the user's manual: When it should be developed and who should receive one.

THE STANDARDS MANUAL

The above outline is a general guide to what standard rules, codes, and methods should be stated before systems development and programming begin. A copy of the standards manual should be available to all personnel concerned with systems development, programming, or use of the system. Each organization will add or subtract from these standards according to the size of the system, types of computer use, and other factors applicable to their particular needs. However, some general ideas about standards documentation can be considered.

Every standards manual should include an introduction explaining how the manual is to be used, how it was developed, and when it should be used. The introduction should contain a table of contents and other references. Since no standards manual is perfect, the second chapter, or section, should outline methods for requesting changes and handling updates. The following section of the manual, which might be called systems designations, will usually explain terminology: the proper names for the different areas of the organization; what systems have been developed or may be developed in the future; the methods for numbering documents; a brief outline of the computer system available; the steps to developing a system; and any other general information to acquaint the nontechnical person with the use of the computer. Following these general chapters, the manual should then become specific in each area of systems development and design.

Plans for systems studies: This section of the standards manual explains the steps to follow to initiate a systems study. In other words, persons in the organization who feel they could make good use of the computer should have some definite, formal method of requesting to use it. This is usually a form or memo filled out and sent to the person responsible for systems development. The manual should also explain approvals for using the computer, how they are made, how systems studies are requested, the steps that will be followed if approval is made of the request for a system study or development, how personnel will be assigned to this system, and perhaps a checklist of things to do or not to do before a system study begins.

Systems design standards: This section contains standards the systems designer should follow for all systems applications. This would include standard codes such as *s* for single, *m* for married, 1 for male, 2 for female, etc.; work sheets which can be used during systems design; specification of the type of input and output layout sheets to be used; decision table formats; standard names to be used, and use of equipment. Available system resources should be listed as well as procedures to handle requests for new equipment, specific formats for listing data elements, and methods of preparing systems flowcharts, etc.

Programming standards: The programmer uses this particular section of the standards manual to determine his methods of problem analysis, the materials he will receive from the systems analysts and designers, and the computer languages he may use. For example, some installations specify that only COBOL can be used as a computer language; therefore, the use of any other language would require special permission. The documentation which the programmer is expected to supply will be listed along with the methods of logical flowcharting, keypunch conventions, methods of coding, testing, input-output design, logic, use of the job control language for the computer, and the use of sorts, merges, and transfers. A section should be included on data communications if the organization is currently using terminals.

Standards for documentation: It may seem to be carrying documentation too far to say that there should be documentation about documentation. However, the standards manual should include a few specific paragraphs about the documentation expected from each person involved in systems development and programming and the methods used to prepare documentation. Many organizations use automatic flowcharting or automatic documentation packages. Some organizations use kits; others use typed documentation. This information should be stated for the personnel who will prepare the system documentation.

Run book preparation standards: This section contains methods which should be followed by either the systems analyst, designer, or programmer for preparing computer run books and operating instructions. Operating instructions will include statements of the programs, systems flowcharts, the detailed operator instructions, use of output equipment, use of input files and mass storage files, types of labels which should be present and kept up to date, how to handle halts and checks, the use of control figures, and distribution of output.

User's manual standards: This last section should list what a user's manual should contain, who has responsibility for preparing it, how it is distributed, and when it is required.

Details on each of the above types of documentation are included in the rest of this chapter.

SYSTEMS DOCUMENTATION

Outlines of the documentation needed for a computer-implemented system are detailed in Figs. 10-1 through 10-6. The documentations needed are listed as:

1. General systems narrative (that documentation needed for management)
2. System study documentation
3. Systems and equipment specifications
4. Program documentation
5. Run books
6. User manuals

GENERAL SYSTEMS NARRATIVE

 1. General Introduction
 2. Problem Narratives
 3. Interview Plans
 4. Inventory Sheets and Library Research Planned
 5. Output Needed Narrative
 6. Input Analysis
 7. Flowcharts or Rough Diagrams
 8. Samples of Documents and Forms
 9. Organization Charts
10. Decision Tables and Other Documented Analysis
11. Study Reports Produced, Usually Interim
12. Present Cost
13. Systems Cost
14. Alternatives in Designs
15. Feasibility Study
     (a) Investment Analysis
     (b) Noncost Advantages and Disadvantages
     (c) Personnel Factors
16. A General Written Recommendation for the Study of the System

---

Fig. 10-1   Contents of the General Systems Narrative (a documentation supplied to management and nontechnical personnel)

GENERAL SYSTEMS NARRATIVES

Systems folder of correspondence: The first general systems documentation should be a file folder of all correspondence about the requests for design of the system. The file contains all letters, memos, etc., which possible users of the computer make to the systems group when they request a systems study.

Management narratives: These are general narratives, usually prepared by the systems project manager about how the system will operate and the time, money, and personnel needed to implement the system.

Organization charts: A complete chart of the entire organization should be included as well as a detailed chart of the area under study.

Network charts: Network charts show the flows within the organization. They are sometimes called work flowcharts and are very similar to systems flowcharting, except they depict flows of raw material, personnel, etc., rather than computer processing.

System description: This is a general management narrative which describes the system in detail and is a forecast or estimate of resources and a statement of objectives. It is not the actual detailed specifications of the system, but a starting point from which systems analysis and design can begin. The system description should define the problem which initiated the systems study, and state the objectives which the study should meet and the objectives of the system operation which will run on the computer. The systems description should also contain a brief resume of the background and qualifications of the personnel selected to make the systems study.

Several narratives should be included in this systems description which briefly list or describe the output requirements. These are essentially the same as the system objectives: General narratives of the input needed to produce the desired output and of storage files needed in the system and currently not present in any other computer system. Systems flowcharts or at least rough diagrams can also be included to give a general idea of how the system might work. The systems description ends with a summary to management of estimations and forecasts, such as equipment, personnel, and other resource requirements and cost estimates for implementing the system. Again, the systems description is an estimate and simply a basis to begin the systems analysis. It may or may not be the same as the final systems documentation and specifications.

Systems analysis and systems study documentation: The systems analysis and systems study documentation may consist of both formal written documentation and file folders of what the systems study team has investigated. As with most documentations, the systems study package begins with a general introduction, probably including the names and qualifications of the project manager and team members. A fairly detailed narrative of the problem and objectives of the systems study should also be included. These are the final objectives and description of the problem, decided *after* the systems study has been made. These may or may not differ from the general systems narrative included in the documentation before the project begins. The systems study is based upon research, interviewing, personnel, and analysis of the system requirements. Therefore, this documentation package will contain reports on interviews, inventory reports on the personnel and equipment already in the system, and any library or other research that may have been made during the study phase. The documentation package will also hold file folders; these will contain the output needed, the input that has been analyzed for the system, and flowcharts and rough diagrams. One systems flowchart should be made for the systems design team.

SYSTEMS STUDY DOCUMENTATION

1. General Introduction to the Systems Study
2. Written Authority to Proceed
3. List of Team Members and Their Qualifications
4. Narrative of the Problem
   (a) Statement of the Problem
   (b) Statement of Objectives
5. Research
   (a) Documents Inventoried
   (b) Publications Researched
   (c) Interviews and Results
   (d) Equipment Inventories
   (e) Consultants Used.
6. Output Worksheets
7. Input Worksheets
8. Flowcharts and Diagrams
9. Samples of Source Documents Analyzed
10. Present Cost Estimate Worksheets
11. Projected Cost Estimate Worksheets
12. Alternatives Analyzed
13. Plan for Presenting to Management

Fig. 10-2 Contents of Systems Study Documentation (papers used during the systems study)

Many miscellaneous documents are included, such as samples of source documents and forms to be produced, organization charts, decision tables, other documented analysis, study reports, a study of the present cost of the system, a forecast of the systems cost, design alternatives, and a feasibility study. This last should present an investment analysis as well as listings of cost and noncost advantages or disadvantages, personnel factors, and other study facets concerned with economic and technical feasibility.

Finally, the systems analysis phase should end with formal recommendations in writing to the management, who must decide whether to go ahead with the system or to declare it unfeasible. Figure 10-2 lists the various inclusions in the systems analysis and study documentation.

SYSTEMS SPECIFICATIONS

The systems specifications package begins with a table of contents, approval sheets, and general narratives. The narrative should include the philosophy behind the design used and state reasons for the particular design *vs* other alternative methods. The systems specifications would begin with the system objective or the outputs to be produced. Data sources for producing these outputs and how the data originates, is prepared, and converted for the system are also given.

Details, work sheets, record charts, file layout sheets, record layout sheets, report layouts, input, card layout sheets should all be included. A systems flowchart should be prepared and labeled with program runs, data files to be used, etc. This systems flowchart is the basic systems flowchart which will be passed on to the

EQUIPMENT AND SYSTEMS SPECIFICATIONS

1. Table of Contents
2. Approval Sheet
3. Narratives
4. Sources of Data, Origin of Data
5. File Layout Sheets
6. Record Layout Sheets
7. Report Layouts
8. Systems Flowcharts
9. Program Runs
10. General Logic of Programs
11. Job Control Card Information
12. Specifications of Programming Standards
13. Summary of Special Procedures, Formulas, Feedbacks and Controls
14. Code Lists
15. Sample Forms and Data
16. Data Communications or Teleprocessing Needed
17. List of Data Elements and Characteristics
18. The Summaries of the Systems Specifications, Equipment Specifications
    (a) General Systems Summary
    (b) Central Processors Specifications
    (c) Peripheral Equipment Specifications
    (d) Auxiliary Equipment and Supplies Needed

Note: In addition to these specifications, personnel to run the system can also be specified at this time, giving job descriptions and other information as to the personnel procedures.

Fig. 10-3 Equipment and Systems Specifications (to be included in detail documentation)

programmers for reference as to where their programs fit into the system. The program runs which will be needed to implement the system are defined. General logic for each program is developed, along with control card information and programming specifications, such as the languages to be used, the standards to be followed, etc.

One sheet of the systems specification package should summarize special procedures, formulas, feedbacks, and the system controls the programmers will use. These controls could be error messages and other special edits and checks which will be programmed into the system. Every system should have a code list which presents the various code meaning used in the system, as well as sample output and input forms and sample data. A list of all the data elements in the system and their characteristics, both input and output, should be made. Any data communications or teleprocessing requirements should be specified with detailed methods of handling these requirements. As with most documentation, the system specification package ends with a summary, detailing for the programmers factors that they should consider as they begin to code the programs for the system.

EQUIPMENT SPECIFICATIONS

The equipment specifications (Fig. 10-3) are needed if new equipment is to be acquired for the system. This is usually called "A Request for Proposal" and should be

---

PROGRAM DOCUMENTATION

1. Statement of What Program Does
2. Statement of Program Name, Codes, Numbers, Input, Where Input Is From, Output, Where Output Goes (This Is a General Narrative)
3. Systems Flowcharts Showing the Program
4. Procedures, Formulas, Decision Totals, Etc.
5. Logic Flowcharts of Detailed Steps Carried Out by the Computer
6. Layout of Input Files, Output Files, and Storage Files
7. Name and Description of Each Data Element
8. Program Source Listings
9. Samples of Output, If Printed
10. Operators' Instructions
11. Accounting Information, Description of Users, Etc.
12. Persons to Contact for Further Information

Fig. 10-4 Contents of Program Documentation (see Appendix on NASA documentation)

detailed enough so that the equipment manufacturers will be able to reply knowledgeably to the request. Equipment specifications begin with a general systems summary of what equipment the system requires and the specifications for each item. Central processors specifications include the speed of processors, memory requirements, expandability, and other such factors.

Peripheral equipment and auxiliary equipment are also specified in detail. Card readers, printers, and other equipment are specified as to rate of speed, cost, compatibility, and the like. In some cases, equipment manufacturers will supply peripherals, or the peripheral suppliers may differ from the central processor supplier. It is usually a good idea for the managers to obtain several bids on equipment and to evaluate these based upon the equipment specifications.

PROGRAM DOCUMENTATION

Whereas the systems specification is a general specification of the entire system, the program documentation (Fig. 10-4) is one set of documentation about one computer run within the system. Program documentation should exist for every computer program that will be used within the system. Figure 10-4 shows what the minimum contents of program documentation should consist of.

The programmer is responsible for supplying final documentation of each of his programs. Program documentation is a long-term rather than short-term benefit. As personnel change in an organization, only program documentation can tell incoming employees precisely what the program should be doing. In many cases when documentation is lacking, programmers will completely rewrite a program rather than try to figure out the existing one.

The programmer normally begins documentation with a general statement of what the program does, then lists the program name, perhaps the program code number, and its call name within the computer system. Samples of any source documents used to prepare program input should be included. In many cases, the output of one program will be input to the next program. In other cases, the output is distributed within the organization.

Systems flowcharts showing the program being documented are needed. The procedures, formulas, and other analysis of procedures used should be presented in the program documentation. The program flowchart details the steps the computer will carry out to process the data for this particular computer run. Chapter 9 presented several examples of program flowcharts.

The layout sheets used by the systems designers should be included in the program documentation, such as card layout forms, printer layout forms, storage file, and record layouts. Also, a name and description of each data element in the program should be included. These are usually extracted from the systems design specifications. Once the program has been tested and is approved, the final source listing of the program should be included. Next to them should be samples of the output the program produces. Finally, there should be a general description of how the computer operator will run the program. This section does not replace the operator's run book, however.

Finally, documentation should give accounting information as to program costs and users, as well as persons who can be contacted for further information about the program. Titles of persons are preferable to names in such listings.

As stated before, program documentation provides long-term benefits to the organization. It is essential to efficient operations. When an organization or system is well documented, it is usually satisfying the user requirements. When a program or system is poorly documented, however, there are usually many problems with the program for which no answer can be found, because there is no documentation about how the program works.

OPERATIONS RUN BOOK

The operations run book is basically the written listing of the physical steps to be carried out in actually operating the program on the computer. There should be one run sheet or run listing for every program in the system. These can be kept in one large book for the system. In many cases, however, they are separated into three or four parts. One part is kept at the control desk so that the librarian can prepare the input-output files from the magnetic tape or magnetic disk libraries. A second part is kept in the distribution section so that the computer output can be properly bound, photocopied, and distributed. The operator may retain only the operating steps needed to run the program and special instructions, such as error handling, etc. Figure 10-5 shows the minimal requirements for documenting an operations run book.

The computer operator needs two basic things to run a program. First, he requires a systems flowchart showing the input-output files of the program, their system name, and where they can be obtained. Second, he needs an operating "setup," which will tell him what magnetic files go on what drives, the name of the run, how to handle errors, what messages he may receive from the program, what type of output he is expected to get, what special forms may be needed, etc. If possible, this information should be contained on one $8\frac{1}{2}$ x 11 sheet. A second sheet will give information on how to label the input-output files which are created by the program; what to do with the output; where it is to be distributed; special procedures for halts, checks, and reruns; the program job control language; and any special supplies which he may need, such as a special carriage control tape, special printed forms, and so on.

RUN BOOK DOCUMENTATION

1. Statement of What Program and System Is to Be Run
2. A Systems Flowchart
3. List of Data Files Used
4. Operating Steps Consisting of Setup, Run, Error Handling, and Messages
5. Special Forms
6. Output Distribution
7. Labeling
8. Halts, Check Points, and Reruns
9. Job Control Language

Fig. 10-5 Information that needs to be documented for computer operators

---

USER'S MANUAL

1. Table of Contents
2. General Systems Description
3. Systems Flowcharts
4. Requirements for Input
5. Input Layouts
6. Output Produced
7. Output That Can Be Requested
8. Output Formats
9. Procedures Used
10. Samples of Forms
11. Methods of Feedback
12. Whom to Contact

Fig. 10-6 Contents of the User's Manual

USER'S MANUAL

After the system has been tested and implemented, the user's manual should be produced. This documentation is written for nontechnical personnel using the system. Nontechnical personnel may include accountants who use the computer to produce general ledger listings and cost accounting reports, salesmen who use the computer to analyze their sales, and engineers or production people who use the computer for bill of materials, bill of lading, shipping, warehousing, etc. The user's manual should describe the entire system and give directions for its use to all personnel who may be involved with the system. Figure 10-6 lists the minimum contents of the typical user's manual.

The systems analyst is the person most likely to prepare the user's manual. It should be written in a clear style and free of detailed, technical descriptions. The introductory sections describe the system overall, why it developed, what it can do for the persons within the organization. The general systems flowchart is presented to show the overall procedures, files, and program runs within the system. The early sections will also describe how input is entered into the system, where it originated, and how the user can enter his own input.

The middle section of the manual should contain a narrative of the output produced by the system, the output formats, and an explanation of any other output that can be obtained by special request. The general philosophy and theory behind the procedures used to produce the output should be included. This is especially true when mathematical or statistical techniques are the basis of producing output. This section should provide samples of forms and methods for the user to send comments about the system. The final portion of the user's manual usually includes a page about whom to contact for further information, how a special presentation about use of this system can be made to users, etc.

The user's manual is the final system document. It should not only explain the system but also explain how it can work for those personnel who use it. Complex systems often are presented in a "seminar for users," and special training sessions are established to explain to personnel how the system will affect them and how they can use it. For example, when a bank automates its teller windows to have direct access to customer account balances, several days of training are given to explain the system to the tellers. Each teller is also given a user's manual. For management reporting systems and other higher level information systems, a general seminar is presented to acquaint personnel with the overall system as well as what it can do for them.

NUMBERING DOCUMENTS

As documentation is prepared within an organization, the documents should be numbered as they are prepared. A typical structure of document numbers could be:

1. Department number
2. Responsibility code
3. Account number
4. Project number

By using these numbers any document can be readily identified in terms of the project and system it belongs to and also who has responsibility for it. For example, Document No. 1AA-320-106 could be identified as follows: 1 refers to the Finance Department; AA is the responsibility of the accounting manager; 320 is the account number to which costs are applied; and 106 is the project number under which the system is being developed. Therefore, before any documentation begins, a method of coding and numbering documents for easy identification should be devised.

SUMMARY

The preparation of adequate documentation begins with proper standards for procedures, rules, methods, and makeup of the actual documentation. Without proper standards, documentation will differ from system to system; codes, names, numbers, etc., will vary from other documentation for the systems within the organization.

The minimum documentation needed by any organization is as follows:

1. Management narrative
2. Systems study and feasibility phase

3. Systems and equipment specifications
4. Program documentation
5. Computer run books
6. User's manuals

The number of copies of documentation will vary according to application and system size. However, one or two copies of the systems request and approval sheets are usually adequate—one to be kept with the project manager, the other with the top management personnel responsible for systems development. Several management narratives about the system may be distributed to the managers and users within the organization. Only a few copies of the systems study and feasibility study are needed; however, the work sheets and file folders which make up the systems study should be available to all persons who participate in the systems development and design. A copy of the systems specifications and the work sheets and folders that accompany them should be available to every member of the systems team as well as every programmer.

Every computer program in the system should be documented individually. Computer program documentation is mostly a long-term benefit and, therefore, is not used for daily reference. However, it should be easily accessible so that any personnel needing information about a particular program can retrieve it. Every system should have at least one complete computer run book. All the computer operators in the operations section should have copies of these run books; usually, at least one or two pages should describe operating procedures for each computer program. Finally, the user's manual should be the most widely distributed documentation within the organization unless, of course, the system is classified in some way. The user's manual should be available not only to top and middle management people concerned with the system, but to all personnel working within the system.

Documentation is the basis for the continuing efficient operation of the system. Because computer programming and systems design are often abstract and use many codes and symbolic meanings, it is impossible for any one person to remember all aspects of a system. Therefore, written documentation is required. Personnel turnover, job changes within the organization, and other personnel factors will mean that the new personnel must have adequate documentation to familiarize themselves with the system. Most management people within the computer industry will probably agree that poor documentation will make the system inefficient and difficult to maintain. On the other hand, good documentation may not necessarily correct poor systems design or poor programming practices, but it will usually contribute to efficient system operation.

Questions

1. Define documentation.
2. What six areas require documentation?
3. Explain the need for a standards manual.
4. What is a run book?
5. What is the user's manual?
6. Who has the responsibility for preparing program documentation?
7. Why should documentation be used?
8. Who uses documentation?

Problems

1. Obtain a computer program and prepare documentation for it, using this chapter and the NASA guide in the appendix as reference.
2. Check with your company or school computer center to obtain samples of documentation. Write a report on the completeness of this documentation.

11
Cost Considerations

The systems analyst is frequently asked not only to determine if the system is technically feasible but also to consider if it is economically feasible. This section presents the economic considerations which the analyst may consider if presented with the problem of studying system costs.

Organizations basically have two transactions which affect them economically. One is incoming revenue; the other is outgoing cost or expenses. The systems analyst is usually not concerned with revenue as an economic consideration but only with cost. Stated simply, the system to be implemented should meet all management objectives at the lowest possible cost. This does not mean that cheapness is the objective of cost planning. Rather, obtaining quality at a reasonable cost is the objective.

Cost can be broken down into six types, described below.

Fixed costs: Fixed costs do not change with volume of production. These are costs such as salaries, equipment rental (unless the equipment is rented hourly), floor space, air conditioning units, and the like, which do not change no matter how many hours the computer is used.

Variable costs: Variable costs are those that change as the rate of production changes. In other words, a computer system which produces 300 reports monthly will have a much higher cost for paper supplies, card punch supplies, etc. than a computer system which produces only 15 reports a month. Therefore, variable costs change as the use of the computer changes.

Controllable costs: Controllable costs are those which can be controlled by efficient procedures and methods. These are the costs the computer systems analyst is interested in studying. Controllable costs are costs such as computer time, which can be controlled by efficient scheduling of the computer, and supply costs, which can be controlled by wise purchasing and inventory control techniques.

Noncontrollable costs: Noncontrollable costs are those which cannot be controlled by the organization. Some examples are taxes, licenses, and other government fees. In most cases, noncontrollable costs do not concern the systems analyst, except to be reported accurately.

One-time costs: One-time costs occur only once in implementing the system; for example, such costs as the systems analysis phase, the systems design phase, programming, data conversion, etc.

Recurring costs: Recurring costs will occur every month and are necessary to keep the system operating. They include salaries, supplies, equipment rental, and other such costs.

One of the first jobs of the systems analyst is to categorize the costs into one of

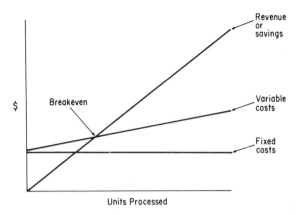

Fig. 11-1 Break-even analysis

the six kinds listed above. Generally, computer systems have high fixed costs and low variable costs; that is, the higher the volume, the lower the unit cost will be. This is true in most manufacturing, retailing, and other businesses. Figure 11-1 charts the fixed costs and variable costs per volume of computer processing. Sometimes it is difficult, nearly impossible, to estimate the cost savings or revenue which using the computer would give. However, if this figure can be estimated, another line would be added to Fig. 11-1. In this case, the revenue line, which is the total of either additional revenue or costs savings, is drawn and a break-even point can be estimated. In Fig. 11-1, the break-even point is where the revenue line and the variable cost line cross. In a business organization, this is similar to the sales line and the variable costs and fixed costs lines.

COST BENEFITS AND COST EFFECTIVENESS

Many economists, writers, theorists, and workers in computer science use the terms "cost benefits" and "cost effectiveness." Definitions vary on the meaning of these terms; however, a cost benefit generally refers to any monetary saving or income arising from some decision or systems design characteristic, such as reduction of personnel, increased sales, elimination of duplication, etc. There are only two ways to obtain a cost benefit; that is, either to increase revenue or to reduce costs. As stated before, whether the organization is a profit or nonprofit organization the systems analyst makes his greatest contribution by reducing costs. Of course, the system must operate efficiently. Many benefits are very difficult to estimate quantitatively, such as the more efficient operation of an organization.

Cost effectiveness also has been defined in various ways. One definition is that the cost returns something valuable to the organization. However, putting a quantitative value upon this asset is difficult. The computer is a modern business technique. Therefore, the question might be: How much is it worth to an organization to be modern?

The dollar figure for the answer to this question frustrates many personnel, not only the systems analyst but top management and financial accountants. At this time there is no real answer to the question of cost estimating. The systems analyst should avoid applying any cost figures which are debatable. Many cost estimates are approximate figures. Any claim to cost savings for such items as better administrative efficiency, increased morale, better customer relations, etc., should be studied very carefully before being included in the systems analyst's report.

COSTS IN A COMPUTER SYSTEM

Although some costs are very difficult to estimate, certain costs can be estimated readily. Although many of these figures will be approximations, the costs usually can be estimated to such a degree of accuracy that a valid economic analysis can be made of a proposed system. The first type of costs to be considered are the one-time costs which will be incurred during systems development.

ONE-TIME COSTS

Transition from the old system to the new: Such costs as hiring computer time from service bureaus, the installation and transportation fees incurred, the additional personnel or overtime that may be worked, etc., can all be considered one-time costs for the transition from the old system to the new.

Conversion costs: Files that are not ready for the computer must be converted. These costs are considered one-time costs because they are performed to convert the initial written records into computer readable media. Preparation costs occur also, as transactions occur while the system operates; however, these costs are different from the one-time conversion costs. Once the data file format and the record formats have been decided upon, data conversion costs can be estimated fairly accurately. The formula presented in Chap. 7 can be used to determine conversion costs for keypunching from source documents into punched cards.

Analysis and design costs: The costs of systems personnel to analyze the system requirements and design the system specifications is a one-time cost which must be included in any economic analysis of implementing a system.

Programming costs: Programming costs are also one-time costs incurred in developing a system. Later in this chapter, detailed breakdowns of both system analysis and design costs and programming costs are presented as well as the methods of obtaining these costs.

Consultants: Usually when a system is implemented, particularly a large one, several consultants will be used not only to get an outside view but also to obtain specialized help in certain areas such as remote communication, mathematics, and other such functions. Consultants can be considered a one-time cost of systems development.

Computer time: An accounting system should be established to differentiate between computer time used for running a system and computer time used for testing or analysis of the system. Therefore, the one-time cost included in the estimate is that of how much computer time will be needed by the programmers to test their programs before the actual implementing of the system can take place.

RECURRING COSTS

The other type of cost important in estimating the expense of operating the system are recurring costs. These costs will recur every week or month as the system operates. They are:

1. Data preparation
2. Personnel
3. Equipment rental or depreciation
4. Power or environment control
5. Supplies
6. Maintenance programming
7. Maintenance of hardware
8. Training
9. Insurance
10. Space
11. Communications
12. Transportation

Data preparation: The cost of preparing data occurs as transactions which affect the system take place within the organization. In some systems, such as a real-time response system (airline reservations system), personnel at remote terminals enter data directly into the system. This usually is considered a personnel cost for remote terminal operators and not a data conversion cost. However, in most systems, data is entered through the computer after being converted from some written form to some computer form. By determining the number of transactions that will occur, the systems analyst can approximate the weekly or monthly cost of data conversion.

Personnel: System personnel are included as a recurring cost. The salaries of those who do nothing but operate the system are total salary costs for the system. Other personnel, such as computer operators, programmers, keypunch operators, etc., may charge only part of their salary to the system. By obtaining estimates of personnel requirements in the systems design phase, the systems analyst can obtain good approximations of personnel costs.

Equipment rental or depreciation: Once equipment has been selected, the systems analyst can predict almost exactly how much equipment will cost for the system. However, where the equipment is shared by many systems, the hours of computer time run must be estimated and a cost figure obtained from this. In multiprogramming or time-sharing situations where the computer is shared by many users, no direct hourly cost is possible. To arrive at an approximate cost, the analyst either estimates the central processor time needed or examines similar systems.

Power and environment control: Computer systems use large amounts of electricity and also must have special environment control, such as large air conditioning units and humidity controllers. These costs may or may not be considered by the systems analyst, according to whether the accounting system of the organization charges these costs directly to the system or includes them in overhead.

Supplies: Computers use quite a few supplies, including paper, carbon ribbons, punched cards, etc. Most supplies are considered as expenses, but disk packs and other such storage media should be considered capital assets and depreciated.

Maintenance programming: Most programming costs are one-time costs. However, some programming, which is necessary to maintain the system, may occur once the system is implemented. When tax rates change, for example, the program must be adjusted to meet these new figures. Sometimes in large systems, several programmers are assigned to maintain the programs in one system. Charges for their time and their computer testing time will be a systems recurring cost.

Maintenance: The computer manufacturer ordinarily charges a monthly maintenance fee to maintain equipment. In some cases, this is a fixed charge; in others, it depends upon how much maintenance is needed, how many parts must be replaced, etc.

Training: All systems which are implemented usually require training not only of the user personnel but of the computer personnel such as operators, librarians, keypunch personnel, etc., who must function within the new system. In large organizations, training officers are full-time and usually have several instructors to prepare presentations of new systems and new techniques. The training is a direct cost to the system and will probably be a recurring cost as the system develops and changes. Also personnel turnover requires additional training and, therefore, additional training cost.

Insurance: Insurance upon computer equipment will be required and may be charged directly to the system or may become part of overhead costs.

Space: The floor space needed for equipment by personnel work areas, etc., may be charged directly to the system or may be carried as overhead. Whatever method is used, the systems analyst should try to obtain some idea of the floor space needed for the system and to include these in his systems cost estimate.

Communications: Systems which use remote devices incur a monthly charge for telephone lines to carry on the data communications. (This aspect of a system is covered more thoroughly in Chap. 12). However, monthly communication line charges should be estimated and included in any economic analysis.

Transportation: In some cases, data is transported from the computer center to remote locations via parcel post, freight, automobile, truck, or some other means. These transportation costs should be included as a monthly charge.

All of the above costs are usually approximated. However, the systems analyst's job is not to *prove* that the system is economically feasible, but to analyze correctly and truthfully *whether* the system is economically feasible. Therefore, estimates which are reduced or omitted will make any economic analysis unreliable. The systems analyst must report accurately to management about the use of the computer. It does not help the organization if costs are estimated so as to make the use of the computer seem more desirable. In most systems in the past, costs have been underestimated, in some cases very much. Therefore, the systems analyst should probably be liberal when estimating costs, especially when estimating manpower and time needed to implement the system.

THE FEASIBILITY STUDY

The three factors involved in feasibility studies are technical feasibility, economic feasibility, and operational feasibility. Technical feasibility is usually easy to determine by a few inquiries to manufacturers. Operational feasibility usually depends upon the quality of the systems analysis design and programming of the system. The basic consideration of most feasibility studies—and of this chapter—is economic feasibility.

Basically, economic feasibility is determining whether the computer implemented system is more economical than the old system. In the majority of cases, the system which is being studied will be replacing an older system, perhaps a manual system or a manually automated system. Therefore, by obtaining cost data on both systems, a reasonable difference between the costs can be determined. Figure 11-2 shows cost data for a typical analysis of system feasibility. Column one shows the old systems cost, and column two shows the new computer system.

Cost Item	Present System (Manual)	Proposed System (Computer)
ONE-TIME COSTS:		
Systems development		5,000
MONTHLY COSTS:		
Personnel	125,000	72,000
Data preparation		16,000
Equipment	14,000	31,000
Utilities	500	1,100
Supplies	8,000	10,000
Maintenance	2,100	3,200
Training	300	2,000
Insurance	950	1,800
Space	6,000	2,000
Communications	800	2,000
Transportation	1,100	1,800
	158,750	142,900

Present system:	$158,750
Proposed system:	142,900
Savings:	$ 15,850

Fig. 11-2 Cost data comparing a manual system with a new computer system

Cost data can be presented as total cost or as unit cost. Where units of production can be determined and are important, the data should probably be presented in unit costs. In other words, it is more important to determine difference in what it costs to produce ten, twenty, or thirty thousand invoices a month than it is to determine the total cost of producing an unknown number of invoices per month.

ANALYSIS OF ECONOMIC RETURN

In most cases, the decision to invest in a computer system or any other investment depends upon various factors and can be analyzed by three specialized techniques. Before these techniques can be used, however, the cost savings or increased revenue must be a positive figure. In other words, if the old system costs less than the new system will cost to meet the same objectives, then there is no need to analyze whether the project is economically feasible, unless management is interested in noncost benefits, such as having the latest techniques, or improving customer relations. The three techniques widely used in financial circles are the *rate of return analysis, payback period analysis,* and *analysis of present value,* which is sometimes called the discounted cash flow method.

Rate of return: To figure the rate of return, the amount of investment is divided into the cost savings or increased revenue which will be obtained from the system. For example, if $1000 a year is returned by the system and the system will cost

$10,000 a year to operate, then the rate of return on investment is 10 percent. As a general rule, an organization should obtain a rate of return equal to the rate of return of bank accounts. In other words, 6 to 8 percent would be the minimum rate of return expected from any investment. Some organizations have certain rates of return which they will accept, and they reject all smaller ones. For example, a 14 percent return may be accepted, whereas a 13 percent return will not. However, the rate of return is usually used only to get a general idea of what the system will present in the way of cost benefits based upon the investment which will be made.

Payback period: A second way to estimate the economic worth of a system is to estimate how long it will take for the system to pay for itself. The normal rate of payback is three to five years. The cash flow which the system will create for the organization each year is divided by the total investment. Therefore, if a new system saves $30,000 a year and costs $90,000 in capital outlay, the system can be said to pay for itself in three years. Normally, the payback is figured only on capital investment of equipment and not on day-to-day operating charges.

Both the rate of return and the payback period are discussed briefly here, because they are not really significant in determining when a computer system is economically feasible. They are used to get a general idea of the system's worth or to rank the system with other company projects. However, probably the only true method of determining a system's worth is to use a present value method, sometimes called a discounted cash flow method.

Present value of money: Money has value, depending upon the time received. A thousand dollars received today is more valuable than a thousand dollars received a year from now, the reason being that the thousand dollars received today could earn money if placed in a savings account for one year. Therefore, the thousand dollars received today is approximately fifty dollars more valuable than a thousand dollars received a year from now. Money received in the future can be discounted in terms of present-day value of money. This is called the discounted cash flow method. It can be used to rank various alternatives in systems design as well as the terms of purchasing the computer, that is, whether the computer should be leased, purchased, or financed by some type of debt arrangement. If cash flow is large in the early years and small in the later years, the system is more costly than if the cash flow is small in the beginning years and greater in the future.

Although the present value method is probably the best method for estimating the desirability of investment in a particular project, it has some drawbacks. It is difficult to determine the percentage rate which should be used to discount future dollars. In other words, should 9 percent, a prevailing bank prime rate, be used as the discount percentage for future projects, or should the rate of return of the organization's current earnings be used? Furthermore, some persons argue that the percentage rate should be after taxes; that is, a 6 percent rate should actually be discounted at 3 percent, which would account for a 50 percent tax rate. Nonprofit organizations and government agencies also have difficulty using the present value method because they acquire money through budgets or fund-raising and they must consider their tax structure is nil.

Considering all methods of estimating investment return or desirability, probably the only truth that can be stated is that if there is wide disparity between various investment returns, then the project with the highest return should be invested in. If the question is whether to invest at all, then a project which returns 15 percent (when a 10 percent investment return as a minimum is desired) is probably feasible. However, where only small amounts of percentages separate the criteria, there is probably no factual basis for either dismissing or accepting a project. For example, if one system returns 12 percent and another systems returns 11.5 percent, it would be very difficult to determine which system was truly the more feasible. The methods of analyzing investment returns are just not that accurate. Also, any method used will be only as accurate as the estimates made. Therefore, the analyst must be very careful both in estimating returns and in considering possible returns or percentages of desirability of the project.

CONTROL OF COSTS

The systems analyst must not only estimate and predict costs but also control them, both in the development and operation of the system. Controlling costs usually involves other areas of an organization, such as the accounting section and management. However, no systems analyst would really want to plan, develop, or operate a system without having adequate methods of both controlling and reporting costs. Therefore, the following controls should be set up so that accurate records and reports about the system are available at all times:

1. Adequate record keeping
2. Cost standards
3. Budgeting
4. Cost accounting and reporting
5. Periodic audits

These five controls not only allow accurate pictures of how the system is developing and how it is operating, but also prove or disprove the systems analyst's estimates, projections, and other economic considerations.

Adequate record keeping: Adequate record keeping is not just accounting information but is rather information to be used for planning and control. Before attempting a systems development or becoming responsible for any systems development, the systems analyst will desire a record keeping system which reports costs by categories. The categories which should be kept are:

Analysis cost: This includes meetings, pre-planning, discussions, interviewing, research, and all other functions of the systems analysis phase.

Design costs: Design costs include preparing record layouts, determining data elements and all other phases of systems design, writing equipment specifications, and any other type of work which will be used in systems design.

Programming costs: Programming costs should be divided up into two categories: personnel time spent at programming and equipment costs to test the program.

Documentation: This cost category involves the time spent preparing documentations for the computer system or programs.

Training time: This cost category includes the time spent either in training sessions or in teaching or supervising training sessions.

Data preparation: Data preparation is the one-time preparation needed to convert the old files into the new computer media files. The cyclic preparation which goes on during system operations is the cost of operating the system and not a cost of systems development.

Systems testing: This includes all time spent testing the system whether by machine, manual observation, or whatever other methods are used.

Using this record system, each person working in the systems development area reports the time spent in each cost category. The number of hours worked, the percentage of those hours worked in each category, multiplied by the pay received for those hours becomes the basic cost of each task performed in each area. (Of course, overhead and other factors such as employer contributions to retirement, etc., must also be figured into this pay.) Once these figures are obtained, they can be applied to each category, and a dollar figure by week, by month, etc., can easily be obtained as to the amount of money spent in each phase and also the percentage of the phase completed. For instance, if it is estimated that $10,000 is needed for programming and $5000 has been spent, the project should be 50 percent completed. If it is not, something is not working as planned. Therefore, these figures provide control for the systems development phase as well as accounting information to be used for planning and responsibility of each phase.

TERMS FOR ACQUIRING COMPUTER EQUIPMENT

An organization may acquire computer system equipment in many different ways. Outright cash purchase of all the equipment is one possibility. Other methods of acquiring computers are leasing, renting by hourly rates, third-party leasing, and debt financing. Management policy will determine how the computer is to be acquired once the decision has been made as to which computer and how much equipment to acquire. However, the systems analyst should be familiar with the terms of acquiring computer equipment and how they might affect decisions on equipment use and other economic considerations.

Purchasing: Purchasing a computer outright is very rare in today's industry. High cost of equipment (Fig. 11-3) and higher returns on money in other places usually negate outright purchase of any major piece of computer equipment. Many supplies such as filing cabinets, possibly some card handling devices, may be purchased outright, but the central processing unit and its peripheral devices are normally not purchased. If a computer is purchased outright, that is the total cost of the computer. In other words, there is no present value discount factor to be figured. So by simply putting off the purchase price one year a substantial amount of money can be saved. In other words, as stated earlier, purchasing the computer today for $100,000 would be more expensive than purchasing the computer today and paying $100,000 for it a year from now, because of the interest that could be earned on the $100,000 in that year. When terms of payment for computer use are spread out over many years, the real cost of the equipment can be discounted even further.

Leasing: Most computer manufacturers are pleased to lease their equipment rather than to sell it outright. In fact, before a Supreme Court ruling determined that IBM Corporation must sell its computers as well as lease or rent them, IBM would not sell any computer equipment outright. Since this ruling, all manufacturers have provided both purchase and lease type contracts. The federal government, which is a very large user of computer equipment, has a standard leasing contract which it normally uses for acquiring computer equipment.

Normally, a lease on computer equipment will run for one to five years. The price per month for the lease decreases as the time increases. The typical lease contract also includes maintenance and usually some other options which may be used in the computer center. Leases vary from manufacturer to manufacturer. However, most of them are limited in many ways. Some leases limit the hours per month of use of the

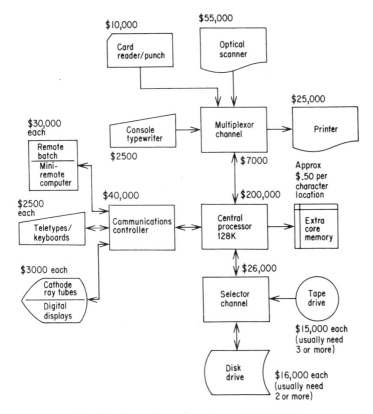

Fig. 11-3 Costs of a medium-size computer system

computer, for example, 150 or 200 hours per month. Any additional hours of use involve an extra payment. Other contracts allow unlimited use for a set price. There may or may not be some penalty for breaking the lease, if the person leasing the computer does not fulfill the entire agreement.

A typical lease agreement may state that for $8000 per month the company will have use for 200 hours per month of one central processing unit of 32K, one card reader, one printer, four magnetic tape drives, and two magnetic disk drives. This is probably an average or typical configuration, and the cost of $8000 per month is probably an average cost. The cost does not include personnel, supplies, etc., but may include maintenance. Manufacturers also differ in what they provide with a lease agreement. Almost all manufacturers will provide a basic operating system for the computer. Some will provide certain application programs such as accounts receivable, accounts payable, and general ledger accounting. Others charge for such programs.

The term "unbundled" has become common in the computer industry. It means that the various components which are purchased or leased with the computer system are not all one price figure but have separate prices. In other words, the computer equipment and operating system have one price, and the software components such as the Sort, COBOL compilers, inventory control programs, and other types of applications programs have different prices. They must be paid for by the organization leasing the equipment or they may not be used. In the past, all of these applications were usually available to a computer purchaser or leaser. Today, however, almost all elements of a computer have separate prices.

Third-party leasing: Third-party leasing is sometimes called sale and lease-back. Essentially, the computer equipment is purchased from a computer manufacturer and then sold to a third party from which it is in turn leased to the organization requiring the computer. The third party is usually a specialist in credit and leasing, such as a bank, finance company, or some experienced equipment-leasing organization. In some cases, third-party leases may be somewhat less expensive than leasing directly from the manufacturer. However, the dependability of the lease contract and the services from the leasing agent may be somewhat questionable, especially if the company is new or just beginning to specialize in leasing computer equipment. Third-party leases do not affect maintenance and other services provided by the manufacturer.

Rent: Renting differs from leasing in that renting usually connotates an hourly rate for computer use. Renting computer time is very common at service centers, where any organization may rent a certain number of hours per month from the computer center. Also, organizations may rent computer time from a neighboring organization in the building, simply using the computer and paying by the hour for the time used. Renting computers may be very complicated, particularly where multiprogramming or remote communications systems are used.

Rent may involve an entire system for one hour or it may be per minutes or seconds of central processing time used, plus the number of storage characters held on tape or disk, plus number of seconds using the channels to transfer data from the central processor to the peripheral devices, etc. Renting time is very economical; however, the organization must be certain they understand the rental terms and the systems analyst should be certain he understands all the costs associated with renting. It is very simple to say that a computer costs $100 per hour; however, when the contract is read closely, it may be that only the central processor unit costs $100 an hour, and channels, peripheral devices, storage space on disk, and other such costs are additional.

Debt financing: A computer system may be acquired through debt financing; that is, a bank loan, a bond issue, or any other method of obtaining equipment on terms or credit. Debt financing usually means that the manufacturer receives cash for the computer as in a purchase, and the organization becomes in debt to some lending agency. However, responsibility to the lending agency naturally varies according to the contract between the organization borrowing the money and the lending institution.

A present-value method can be used to determine the best method for acquiring the computer necessary for implementing the system. The systems analyst who becomes involved in a lease *vs* purchase decision should study the discounted cash-flow method carefully.

ECONOMIC PRINCIPLES

The economics of computer systems probably do not vary significantly from the economics of other types of equipment and systems acquisition which organizations employ. A very thorough review of the economic principles and considerations which affect the selection, analysis, and use of computer equipment and systems can be found in *The Economics of Computers* by William F. Sharpe. The book is published by Columbia University Press and is available in most libraries.

The following four principles are necessary considerations in analyzing any economics of computers.

Lowest cost is not always the best: The most inexpensive piece of equipment or service which can be acquired is frequently not the best or least costly in the computer industry. For instance, although an inexperienced systems analyst or programmer may receive a salary much lower than the average analyst or programmer, his inexperience may prove more harmful and costly in the long run than any amount saved from his salary. This is true of almost all personnel who work in a computer environment. The same is practically true for equipment, although electronics technology creates many exceptions to this. For two pieces of equipment whose specifications are exact and which are guaranteed by the manufacturer, the lowest cost is probably the best. However, rarely do two pieces of equipment, even if they both accomplish the same task, have the exact same specifications. On peripheral devices, data transfer rates will vary as will the storage capacity on magnetic tape, density, etc. Also, the software for peripheral devices must be considered, for even though the peripheral device may be inexpensive, the software may be costly. The principle also applies to supplies, such as cards, punched cards, magnetic tape, magnetic disk, and almost any other equipment or service within the computer industry.

Accomplishment is more important than cost: Many computer industry personnel use a term known as "cost effectiveness." Cost effectiveness essentially means that a certain amount of accomplishment is obtained for a certain amount of cost. If the equipment does not perform the task it is needed for, it will not be effective. Therefore, the performance of a piece of equipment, a person, a supply, a program or a systems design is much more important than how much that particular item or person will cost. Therefore, the systems analyst usually concentrates on accomplishment first and cost second. Cost, however, puts constraints on all areas of systems design. For example, an airline which tried to run its reservation system using sequential tape drives would soon find that it had no customers, because of the time lag between the time the customer requested the reservation and the time he actually found out he had a reservation. Although magnetic tape is cheaper than magnetic disk it is not effective in this case and does not accomplish the real objective, which is to serve the customer in a reasonable amount of time. It is therefore an important economic principle to consider accomplishment before cost. Cost is just a constraint upon the accomplishment.

Most decisions are trade-offs: All decisions concerning the computer use are really trade-offs between using a computer and investing in some other system. Therefore, the systems analyst cannot truthfully state that acquiring a computer system for $30,000 a month would be the best decision for the organization unless he realizes that many other decisions which can be made within the organization may be just as beneficial. For example, a manufacturer may have to decide whether to install a computer system which will provide better management information and clerical administration or a new assembly line which may increase production by 20 or 30 percent. As a specialist in the computer area, the systems analyst is not expected to analyze the entire firm or organization completely. However, he will want to avoid being totally computer oriented. The decision of whether a task should be performed manually or by computer is a trade-off between a manual system and an automated system. Trade-off decisions are probably the hardest decisions any manager, committee, or systems group has to make. Something must be given up for something else to be gained.

When in doubt, use historical or industry guidelines: When managers of computer systems or systems analysts are undecided about equipment, cost, or the results, the best answer may be found in either the many industry guides published within the computer industry or the historical data accumulated by the firm or organization itself. For example, a manufacturing company with 500 employees and a sales of $20 million a year may be confused about how much money it should spend on computers. By turning to industry guides, the company can obtain the typical average expenditure for computer systems equipment for a manufacturing company of its size. Salaries for systems analysts, programmers, computer operators and other personnel also can be obtained from the many industry salary guides published. Types of equipment used are usually readily available in trade magazines or other publications. The federal government offers help in this area, since the government publishes inventories of equipment, guidelines, typical contracts, history of use of computers, etc. Appendix C of this book contains the federal government specifications for standards in information processing applicable to systems analysis. Many other government publications present the economic considerations which must be made when selecting, analyzing, or using computers.

QUALITATIVE SYSTEMS BENEFITS

The previous section has presented the quantitative systems benefits in costs savings or added revenue. However, many other benefits accrue from installing computer equipment. The list below is meant only to stimulate ideas which a systems analyst or other managers can use to reinforce their own arguments for installing computer equipment. Although none of these benefits are guaranteed, a well-designed and well-managed computer system will probably produce them. IBM once had a slogan that "machines should work and people should think." Perhaps this benefit alone is enough reason to install more automated equipment. The benefits are as follows:

1. Personnel costs diminish
2. Inventory levels are optimized
3. Deadlines are met and production is improved
4. More and better information for decision making and planning
5. Improved accuracy
6. Space savings
7. Improved communications
8. Freeing of personnel for more productive tasks
9. Time savings
10. Improved utilization of resources
11. Improved customer relations
12. More control of the organization
13. Centralized organization
14. Profit center identity
15. Equipment savings

It should be remembered that benefits normally will progress with time; therefore, at the outset of automation, the organization will more than likely realize clerical benefits, whereas later, better management information becomes one of the most important and valuable benefits.

The benefits listed are, in many cases, self-explanatory. Personnel costs usually diminish in the long-run rather than in the short-run; that is, a computer system can handle greater volume than any set number of persons can. In fact, for a computer to process 40,000 customer statements rather than 20,000 makes very little difference in cost or processing time. Only three or four extra hours of computer time may be needed. However, for manual clerks to process twice as many invoices of that volume, several additional personnel and much overtime might be involved.

The benefit of optimizing inventory levels is not just a quantitative benefit in the sense that money is saved by carrying optimum inventories, but rather that efficiency is promoted by parts, supplies, etc., always being available in inventory, production is more efficient, and deadlines are met.

Another benefit is the capability of the computer to process large volumes, enabling the organization to meet deadlines that it once was unable to meet.

One of the most valuable benefits of the computer system is making more information available for decision making and planning. Of course, as noted in Chap. 6, too much information or information that cannot be readily assimilated is probably as bad as too little information. However, information once unavailable, such as daily sales totals and inventories, can help the manager make better decisions and more efficient plans. The computer also promotes accuracy among personnel who use computer output or prepare computer input. The many editing techniques and other checks available in computer processing encourage personnel to prepare accurate information for the computer because if they do not, it is fed back to them. Also, the output can be trusted in a well-designed system.

The computer's ability to store large quantities of data in small media, such as 20 million characters on a reel of tape, and its ability to replace more voluminous types of operations reduce filing and floor space. For example, the computer, housed in one room, can produce more clerical work, such as invoicing or payroll checks, than a large room of 40 or 50 clerks. As with inventory levels, saving space is not only a quantitative consideration. The computer's ability to analyze exceptions from normal conditions and to improve record keeping and cost accounting operations can improve utilization of resources drastically. Given the proper systems design, a computer could check 100,000 items in inventory in a few hours and determine which of those were not utilized. A manual system would require many days.

Although the computer has caused its share of customer complaints because of error design and poor programming, in many cases, computers will cause much better customer relations. The computer not only gives the customer more and better information about his status with the organization but it also provides quicker feedback and faster communications within the organization.

Control of an organization depends upon management policies, standards, and objectives. However, the ability to carry out these policies has sometimes been impossible because of an organization's complexity and far-reaching operations. The uses of the computer are limited only by the imaginations of the managers, systems analysts, and designers. An organization can improve its control using the computer's ability to speedily process large volumes of information. Centralizing an organization also increases control, for data can be accessed from remote points through telecommunications. Also, the computer's ability to use codes to identify organization transactions allows for profit-center identity. This permits management to identify the profitibility of various functions within the organization.

SUMMARY

Because the systems analyst advises management on the feasibility, selection, and economic benefits of computer systems, he must be aware of the economic considerations of system implementation and development. The systems analyst is usually more concerned with qualitative or cost savings benefits than with additional revenue that installing a system may bring.

Computer systems have high fixed costs and low variable costs. Therefore, as volume increases, the cost per unit of processing decreases. Cost of processing may be separated into many different categories: among them are fixed costs, variable costs, one-time costs, recurring costs, controllable costs, and noncontrollable costs. A cost benefit is any saving or income arising from a decision or systems design characteristic, such as reduction of personnel, increased sales, elimination of duplication, and increased volume. There are only two ways to obtain a cost benefit—by reducing costs or by increasing income. The systems analysis is basically concerned with two general costs: those of developing and installing a system and those of continually operating the system. Normally, the analyst will look for cost savings in both areas.

The systems analyst's first task is to determine if the system is feasible. Three areas of feasibility are technical, economic, and operational. The analyst is concerned primarily with economic feasibility.

In considering the economic feasibility of a system, the systems analyst will probably first obtain all the cost data of the current system to compare with estimated cost data of the new system or procedure. Once the estimated cost data has been gathered, the analyst may use one of three methods to analyze his figures—the payback method, the percentage of return on investment method, and the present value or discounted cash-flow method. The desirability of investing in a project can be determined by using these three methods, although none of them is fool-proof. The analyst must also decide whether the equipment will be purchased, leased, rented, or financed. The discounted cash-flow method is the best method for making this decision.

Although details of the economic considerations for the entire firm are not the analyst's job, he should be familiar with how a decision to install computer equipment affects the firm. Most decisions in the area of computers will be trade-offs with other areas within the organization. Also, system performance is much more important than system cost. However, cost is a restraint upon any system operation. If the analyst needs more detailed information or a starting point for his economic analysis, historical and industry guidelines are widely available and should be used freely.

Many benefits of the system are not strictly quantitative but are qualitative, such as freeing personnel for more productive tasks.

A final consideration is that the benefits obtained by installing computer equipment depend very greatly upon the sophistication of the organization. An organization which is just beginning to automate from manual clerical methods cannot expect to immediately obtain the benefits of a total management information system. Clerical, administrative, and processing benefits are usually the first to be realized. As the organization progresses over time and becomes more automated and sophisticated, other benefits can be realized, such as better management reporting and more information for decision making and planning.

Questions

1. What is the difference between fixed and variable costs?
2. What is a cost benefit?
3. Name and discuss six costs in a computer system
4. What is meant by feasibility?
5. What is the rate of return?
6. What is the present value method?
7. Discuss some methods of controlling costs.

12

The Use of
Telecommunications

The use of long distance telephone lines by computers is said to be greater in the last few years than the use of long distance telephone lines by human beings. This fact illustrates how important it is for every person in the computer industry to learn the techniques of using the computer by "telecommunications."

Telecommunications is defined as any communication using a common carrier line, such as the Bell telephone system or Western Union, for transmitting data from one location to another. Today, output devices, input devices, and central processors can be linked together from any distant point, no matter now remote the devices are from each other. Every person involved with computers will be affected by the impact of long-distance remote communications. By using telecommunications, business can now keep up with branch operations, transmit data that affects its business to remote points, receive data from remote points, and be instantly informed about any situations affecting its profitability.

Government agencies—state, federal and local—have found telecommunications a great help in all areas of government service. Police departments use computers to inquire about criminals, keep up with automobile licenses, receive traffic reports, and so on.

The use of the computer with telecommunications has had enormous impact on the whole society. New power is given to any person who wishes to expand his knowledge about almost any field. Lawyers may ask a remote computer about research into cases they are preparing. Doctors may input a patient's symptoms and receive diagnoses from a remote computer. Consumers may use computers to calculate income tax, prepare kitchen recipes, and even analyze betting on horseraces, all from their home to a computer at a remote point.

This chapter surveys the use of computers in conjunction with telecommunications capabilities and presents the services available today and the considerations the systems analyst must make to complete his systems design.

Basically, the systems analyst must understand what is available in data communications and what hardware and software can be used with telecommunications in order to develop telecommunications within his organization and standards for using telecommunications in a greater network of systems. The systems analyst should use remote communications if it is efficient and economical.

A TRANSMISSION SYSTEM

Figure 12-1 shows a typical computer transmission system. Basically, there is a sender, a data set, a link, another data set, and a receiver. The data is sent across the link, which can be telephone wires, microwaves, or any other means of communication. Data is sent in bit streams. Codes are used to determine what has been sent across the line.

There are three important types of telecommunications, whose merits should be considered separately. They are:

1. Data transmission
2. Data collection
3. Data communication

Telecommunications has various names, among them teleprocessing, data communications, remote processing, etc.; however, to the systems designer it is one of the three categories mentioned above.

Data transmission: Data transmission is sending data, usually in batches, to a remote location. For example, a remote warehouse may keypunch all of its daily shipments data, collect it into a stack of cards or a reel of punched paper tape, and send it in one batch to a central computer center. At the center, the data may be received on another peripheral device, such as magnetic tape or punched paper tape, or it may be used directly by the central processing unit for processing. There is no interaction between the two devices, other than for error correction, and the same transmission could have been made through some other means such as mail, freight service, etc., except, of course, the data transmission is much faster.

Data collection: Data collection refers to data that is collected from remote points which are polled by the central computer to determine if data is ready to be sent. An example of data collection would be badge readers located in large plants. If there were ten badge readers, the computer could poll each badge reader and collect the data as it was entered into the device. Also, cash registers and other input devices may be connected to a computer, and the data can be sent as transactions occur.

Data communications: This term denotes interaction between the sending and receiving points in a remote transmission. Data communications can be the remote terminal which is receiving output and immediately sending back input, or it may be some type of time-sharing system where the terminals are connected to the resources of the computer, such as the COBOL and FORTRAN compilers or special conversational languages.

Each transmission system has merits in terms of cost and efficiency, and each should be considered as a possible part of a systems design in almost all modern systems which are being implemented on the computer.

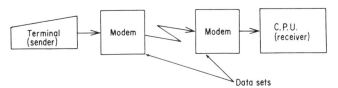

Fig. 12-1 A transmission system

THE SELECTION OF TELECOMMUNICATIONS AS PART OF A SYSTEM

To determine when telecommunications may be integrated as part of the systems design, the following four criteria should be used:

1. Cost of transmitting data
2. Cost of data preparation
3. Need for real-time, immediate response
4. A large number of inquiries into the system

Cost of transmitting data: When transmitting data to the computer center becomes inefficient and costly and where time is lost in transmitting the data through the mail or other means, then using remote terminals to transmit data to the central computer center should be evaluated. Cost of transmitting data is not simply the cost of getting the data from one point to another, but also the costs which may be incurred by not processing transactions in time. Sometimes it is difficult to obtain accurate figures for cost of transmitting data; however, where approximate figures can be obtained, they should be evaluated against the cost of transmitting the data by use of a remote terminal to the central processing unit.

Cost of data preparation: Almost all data which enters the computer system must be prepared so that the computer device can read it. This could include keypunching, key tape, punched paper tape, or any other of the preparation methods discussed earlier. The cost of preparing this data in terms of both the actual cost of preparation and the time lost in waiting for its preparation will determine whether collection devices may be needed or could be used in the system. For example, it may be less costly to connect retail department store cash registers directly into a computer device by remote transmission than it would be to receive the sales slips from the clerks at the end of each day and keypunch them in cards for use by the computer. When the cost of keypunching becomes as large as the cost of the telephone lines and the terminals needed to capture the data at its source, then certainly some evaluation of telecommunications is in order.

Need for real-time, immediate response available from the computer: Airline-hotel-motel reservations, stock quotations, inquiries by a highway patrolman into a status of a license number on an automobile, and the like are examples of situations where a quick or immediate response is essential. Therefore, telecommunications can be used to place the inquirer in direct contact with the computer file to request whatever information is needed.

A large number of inquiries into the system: Where a system has a frequent number of inquiries which could best be handled by direct contact with the computer program itself, then telecommunications is usually needed. For example, where branch offices must constantly ask the home office whether a product can be delivered, where there are constant inquiries into credit status of individuals, and other such situations, a direct terminal to the computer files is advisable. Although immediate response may not be critical, the large number of inquiries that must be handled by the computer system could probably be handled more efficiently by telecommunications than they could be by some type of batch processing.

TELECOMMUNICATIONS SYSTEM DESIGN CONSIDERATIONS

Designing telecommunication systems requires several design considerations which are not typical of systems which do not use remote communications. However, once the communication is completed, data can be processed using the same techniques, same methods, and in many cases the same languages used in ordinary processing. The following considerations are important to the systems analyst as telecommunications is integrated into the systems application:

1. Input-output
2. Software
3. Communication channels
4. Interfaces
5. Volumes
6. Overload and backup requirements
7. Security
8. Terminal capabilities
9. Priorities of use

Input-output: The basic consideration of input-output is that all data formats used in the telecommunications for a particular application be standardized. Also, the codes acceptable to both sending and receiving devices, the bit structure of the codes, the number of characters in a data item, and other techniques must be standardized. Without complete standardization of all input and output, the problems of remote communications can multiply tremendously. A user cannot simply inquire into a computer system from a remote point by typing "I want to know about this particular data item." Conversational programming is possible; however, for most common remote communications some standard formats and standard codes must be devised and integrated into the systems design.

Software: In considering software for telecommunications, the designer must evaluate what will work best for the system and if the software will work at all. For example, COBOL, FORTRAN, and other higher level languages do not yet have remote communication possibilities. Therefore, most programming for telecommunication lines is coded in assembly language, the manufacturer's language for the computer. Many manufacturers provide macro-instructions which can be used to receive input from and produce output to remote terminals. However, the systems designer who plans on using COBOL, FORTRAN, or other higher level languages in his system will have to make some adjustments to the software before it can be used with the remote devices.

The answer to what will work the best is probably that which the manufacturer has developed himself. Programming for telecommunications lines is very complex because each bit must be captured as it is sent across the channel and each character must be handled depending upon whether it is a data character, acknowledgement character, or another telecommunications code. The programming knowledge needed to accomplish all these tasks is specialized, and the programmers developing the user application may not have the needed skill. Therefore, the software developed by the highly skilled, experienced, programming teams employed by the manufacturer will probably work better than any other type of software. By using these communication software packages, the applications programmers can readily use remote communications.

Communications channels: There are many equipment specifications that must be considered for remote communications. A data set or modem (modulator, demodulator) is always required except for audio-type systems. Lines must be synchronized and codes must be understood by all members of the systems design team who participate in the telecommunications design. Most systems have standard start-stop acknowledgements and acknowledgement codes which must be sent as the data travels across the lines or as it has completed its travel across the communications link. Selecting the type of communications channel for the system also must be considered carefully and is discussed later in this chapter. For complete information on available services, the systems analyst should contact the common carrier, either the Bell system or Western Union.

Interfaces: The connection of two devices is called an interface. Connections of remote terminals to central processors is essentially a problem of interface. This interface is made possible by many components which allow the data to be transmitted. The types of connection which are made at various points, the amplitude, levels of the line, frequency, band lengths, and other considerations must be made before there is any decision to use the communications line. In other words, before any terminal or other device is planned to be connected to a central computer, there must be some assurance that an interface is possible.

Volumes: The volume of data to be transmitted is analyzed from both the capabilities of the terminal and the processors. Before specifying any particular telecommunications devices, the systems analyst must consider the number of terminals a central processor can handle, the amount of data which may be entered into the terminals, the speed of the terminal, etc.

Overload and backup requirements: Communications to remote points are frequently slack at certain times and build up at other times. The systems designer will need to determine the overload and peak use of the terminals and also what effect this may have upon the central processing unit. Overloads refer to more data being transmitted than can be handled by either the central processor or the terminal device. Backup requirements refer to what equipment can be held in reserve in case of some malfunction in the main equipment. In many time-sharing and data collection systems, where large amounts of terminals are served, the central processing unit is switched over to another central processor in case of unit malfunctions. Such backup requirements may be essential; however, they may also be very costly.

Security: Where many remote terminal devices access centralized data files of a computer system, some type of security may be necessary so that unauthorized persons do not access data files to which they have no authority. One method used in both remote communications and computer systems is the password. The computer is able to check passwords. Where security is critical, several passwords can be used and they can be updated hourly, weekly, daily, or as required to assure security. However, in reality, no complete security of a data base or data bank has ever been proven. Several computer manufacturers have undertaken studies to determine the best methods of securing centralized data files. However, at the time of this writing, there is no research which proves that data files are completely safe from unauthorized access, especially if the unauthorized access is being made by someone familiar with the operation of computer equipment and software. Other security measures that may be used at remote sites and at the central computer center are the restriction of personnel from equipment areas, use of words which only the personnel would know, such as their mother's maiden name, and other such techniques which can be used not only to protect data files, but to protect equipment and information in all types of situations.

Terminal capabilities: The number of terminals that can be handled and the type of terminals that can be used are important considerations of remote communications. Of course, most of this depends upon the terminal itself, the software which is available for use, and interfaces available. There is no easy answer to determining how many terminals can be handled because this usually depends upon volume of transmission and how many terminals are in use at any one particular time. A computer can be dialed for use just as any phone can. When more calls come in than the switchboard can handle, a busy signal is given. Similarly, when more users call or dial the computer than it can handle, a busy signal is given.

Priorities of use: Any time remote communications are being analyzed, the systems designer should consider what priority will be given for using remote terminals. Since the computer system depends upon the interrupt capability to handle priorities, then the number of priorities that can be handled depends upon the type of interrupts which can be given and the handling of priority interrupts. However, software can also handle priorities based upon code numbers. For example, a large business with many terminals may give a top priority to the executives of the business, whereas the clerical workers in a warehouse may have low priority. In such cases, use of the computer may be interrupted by a higher priority. These situations must be analyzed and planned by the systems team to determine what capabilities are needed in the assignment of priorities.

UNDERSTANDING TELECOMMUNICATIONS

Any communication system, whether electronic or otherwise, consists basically of three components: the sender, the link, and the receiver. Any interruption or error in the communications is usually called "noise." Figure 12-2 shows a typical computer system in diagram form. Most telecommunications systems use the queued method to store both input and output data, so that the central processor or the communications controller can send and receive data according to their speed of operation. Basic components of this system are the terminal, the data set (modem), communications controller, a queue, and the central processor itself. Data is entered into the terminal, goes to the data set where it is transmitted across the data lines to another data set which transfers the data to a communications controller; the controller sends the data to the central processing unit where it is placed in a queue. The central processing unit then processes the data depending upon the priorities specified and places the output back into a queue. Again, according to the waiting line, the data is sent to the communications controller, back to the data set, and across the transmission lines, to the remote data set where it is demodulated to the terminal. Several variations of this diagram can be used in designing teleprocessing, especially depending upon the type of data transmission, data collection, or data communications desired.

Speed of a telecommunications line is measured in terms of a unit called the "baud," which is essentially one bit per second.

TRANSMISSION CHANNELS

The types of transmission channels available are the simplex, the half-duplex and the full-duplex. The simplex channel transmits data in only one direction. The half-duplex can transmit in two directions, but in only one direction at a time. The full-

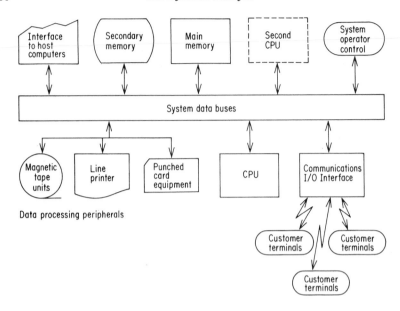

Fig. 12-2 EMR communications system

duplex can transmit data in both directions simultaneously. The channel selected depends on the needs of the system. For example, the simplex line would suffice for simple data transmission. The half-duplex line is practical where received data is returned to the remote points. Where conversational types of communications are needed, such as time-sharing, real-time reservations systems, etc., the full-duplex line may be required.

Before any decision is made as to what services the system will require, the common carrier representative should be contacted.

TERMINALS

The terminal is any device connected to the central processing unit by data transmission lines. Terminals may be:

1. Another computer
2. A peripheral device, such as card reader or printer
3. A typewriterlike device
4. Special equipment, such as badge readers and cash registers

Essentially any device which can be used for input and output can become a remote terminal, depending only upon cost restraints. However, many devices are adapted more readily for use as remote terminals than others. The central processing unit of most modern computers can be used to communicate with the central processing unit or a remote device of another computer. Under program control computers have the ability to dial telephone numbers, connect themselves with other computers, and transmit data. The transmission of data under computer program control is commonly called computer "talking," although no real communication goes on in the human sense.

Specialized devices or any standard peripheral unit can be used as a remote terminal. Card readers, printers, magnetic tape, disk, cathode ray tubes, and other such devices have been adapted for remote communication. However, for data transmission, high speed devices such as magnetic tape are normally used. For data collection, special equipment or punched paper tape and cards may be used.

The typewriterlike device has become a basic terminal in data communications. The operator at the remote point types in the data inquiry or request, which is transmitted to the central processor. The computer returns the output and activates the keyboard which types the answer at the remote point. Typewriters can be simple devices connected to a telephone by placing the phone receiver into a microphone-type holder; the audio tones of the telephone are used to transmit the data. Other typewriters may be more sophisticated and contain sprockets for holding special forms, and may even perform such tasks as typing out invoices and other business transactions at remote terminals.

Much special equipment has been developed to act as remote terminals to computer systems, such as badge readers, cash registers, and similar devices. However, many standard pieces of office equipment, such as the adding machine and microfilm equipment, have been successfully connected to computers. In the very near future, it will be common for computers to analyze pictures, words, and perhaps even voices from remote terminals.

THE CENTRALIZATION OF FUNCTIONS

With the increasing use of remote terminals, more and more functions will be centralized at one computer location. Businesses have already begun to centralize most of their files at one location, where all data may be accessed remotely when needed. In the near future, all data may be stored in large storage files, or data banks. Many business and governmental functions, such as invoicing, shipping orders, reporting, tax billing, subscription renewal, and the like, will be centralized at one location. Any information needed about the data stored in this location can be accessed remotely by the terminal, and data can be selected for analysis as transactions occur. Probably the greatest impact remote communications will have upon society is that all transactions will be processed by the computer as they occur. Therefore, when a customer purchases an item in a department store, not only will the inventory and other accounting functions be updated as the purchase is made, but the individual's credit, checking account, credit card balance, and other such items will be checked at the same time. In fact, it is possible that one day a large data bank will be established where all persons will have one central file and all of the transactions that they make throughout their lives will be filed, monitored, calculated, and distributed to remote terminals by a computer system. The uses and misuses of large data banks such as this are not the subject of this book. However, the implication of a great change in society is apparent when data banks become more organized and more widely used. With intelligent planning and control procedures, the power, efficiency, and speed of the computer may be used for mankind's benefit.

A NEW AGE

Historians have commonly labeled various periods of history by some name, such as the Stone Age, the Bronze Age, the Dark Ages, and so on. Perhaps the age of the future is the Age of Communications. Although the world today has vast communications networks, there has been little progress compared to what will be possible in the future. Some foresee a "public utility of computer services" as common as light, gas, water, and telephone are today, where information is available for everyone. If a terminal can be made economically available to anyone who desires to use it, the potential for communication and for obtaining information is limited only by the imagination. Doctors may inquire into large data banks as to possible diagnosis of patients' illnesses, research in the medical field, and other areas. Lawyers will be able to determine what decisions have been made by what judges at any previous time. Businessmen can obtain information not only of the status of their own business, such as inventory levels, amount of sales, personnel turnover, etc., but also can readily obtain data about the state of the economy, recent stock or commodity prices, or any other such data. Government agencies will have information available as to how to best serve the public. The average man will be able to select information from computer data banks on any subject in which he has interest. Local news, books, schedules of baseball and football games, entertainment, obtaining of tickets for plays, hunting and fishing information, vacation planning, menu planning, and almost any information which can possibly be imagined can be stored in data banks and made available to any persons who have the means—that is, a terminal—to access the central computer file.

SUMMARY

The technology exists today to link almost any input or output device to any computer, including another central processing unit, no matter how remote its location in the world. This capability is important not only to those in the computer industry but also to society as a whole. The use of telecommunications will increase in business, government, and in practically all organizations.

The systems analyst needs to understand what is available in the computer market today, to know what hardware and software can be used with telecommunications, and to develop the specifications and standards which will be used in the application system.

Basically, there are three types of telecommunications in use today. They are data transmission, where data is sent to or received from a remote point, usually in batches; data collection, where many remote terminals constantly collect data and send it to one centralized center; and, probably most important, data communications, which is the interchange of information between two remote devices, a two-way communication.

When considering telecommunications as an intricate part of the system, the systems team must carefully analyze the input-output formats, determine equally carefully what software will be used with the system, and evaluate the capability of the communications channels. Interfacing remote devices depends upon many factors, which also must be analyzed. The volume of data to be transmitted is important both in terms of cost and capability. There must be some provision for system overload or malfunctioning as well as for security, when data is confidential. Before any decision is made to integrate the telecommunications capability into the system, user priorities for the computer must be determined.

Both the Bell telephone system and Western Union provide many services for transmitting data to remote points. Since these services change quite frequently, they should be checked with the organization supplying the common carrier service before the systems design proceeds.

The remote device is called a terminal. It is usually a keyboard type of device. However, many specialized devices have been developed for remote communications, including such common devices as retail store cash registers.

The decision to integrate remote communications into a system depends upon how much it costs to transmit data, what it costs to prepare data, the need for a real-time or an immediate response system, and the number of inquiries which may be made into the system.

The future holds a new Age of Communication, where information will be instantly available and transactions will be instantly processed as they occur. Just as with any new technology, the use of that technology depends upon intelligent planning and effective control measures.

Questions

1. Define telecommunications.
2. What are some specific areas of knowledge which the systems analyst needs to know about telecommunications?
3. What makes up a transmission system?
4. What is the difference between data transmission and data communications?
5. What is a modem?
6. What is an interface? Why is it important?
7. Explain the differences in types of transmission channels.
8. What is the concept of the computer as a public utility?
9. What is a common carrier?

Problems

1. Draw a systems flowchart of a computer that sends data to another computer. Include lines and modems or data sets.
2. Draw a systems flowchart of a computer that is connected to 10 cash registers at remote points.
3. Draw a systems flowchart of a hotel reservation system using typewriter terminals at each hotel.

13

Implementing a System

Once a system has been designed, management becomes responsible for its installation and its efficient operation. In many cases, the computer center manager will be assigned this responsibility; although it could be assigned to the systems analyst. In either case, the analyst is usually involved in establishing the system either as a consultant or as an instructor of personnel who will use the system. The following material explains the activities needed (Fig. 13-1) to effectively install a new system.

Ordering equipment: Any new equipment which will be needed should be ordered and delivery dates set with the supplier. This could include the purchase or leasing of an entire new computer system or simply the ordering of a few additional disk drives. The systems team, which has written the equipment specifications, should work closely with the purchasing or contracting officer who will actually place the order. Almost all companies or organizations that are large enough to use a computer system will have definite policies on purchasing equipment, and the systems analyst will be expected to adjust to these policies. For example, the analyst may specify a certain brand of equipment, but purchasing policies may require three bids on the type of equipment needed. At this time the analyst should also request the contracting of any consultants, teachers, or systems engineers that may be needed from the equipment supplier.

Selection of personnel: New personnel required by the system should be interviewed and selected at a time that will make them available when needed to run the system. Many times it is proper to train typists, file clerks, etc., (whose jobs may be eliminated by the installation of the system) to become computer operators, control clerks, or tape and disk librarians. In many cases, personnel who have previously performed clerical duties can be trained to be computer programmers and even systems analysts. There are aptitude tests available which will give some indication of what personnel are suited for work with the computer system. Experienced computer programmers and computer operators may also need to be hired at this time if they are not already available within the organization.

Education: The training of personnel usually can be classified into two basic categories, technical and user. Technical training includes courses for those personnel who will actually work in the computer center and perform the jobs of keyboarding, operating, programming, report distribution, etc. User training includes the courses or seminars for those persons who either gather the data to be used by the system or use the information produced by the system.

Both types of training should be well planned by the systems team. Course lists and schedules should be made and instructors selected. Personnel who are to attend

Order system, support services (and disk packs)	Order forms, supplies and accessories
Give aptitude tests	Order program products
Select DP staff	Finalize file conversion plans
Education (list each course)	Operator training (may have before and/or after installation)
Document current procedures	Run books documentation
General systems design (workflow)	(operating procedures)
Physical site planning	FE review of physical site
(FE check if hazardous/contaminated environment is suspected)	In-house education
Common carrier arrangements	(orientation of other personnel who may be affected by DP input requirements and output)
Mgmt. progress reviews (recurring)	
Select program products	Install data recorder or keypunch
Select field developed programs	(for early file conversion if planned)
Select A. C. S. offerings	Disk pack delivery
(Application Customizer Service)	File conversion — (application)
*Detail systems design	System test (complete processing cycle)
Order field developed programs	(before installation if possible)
Complete A. C. S. input — (application)	Install system
[repeat for each application]	(Additional) Operator training
Order system disk pack (PID)	
Preliminary file conversion plans	File conversion (if not used earlier)
*Application program development	Parallel or pilot operation
(allow for coding, testing, debugging, and documentation)	Cut-over
Confirm system delivery	System evaluation

* Including A. C. S. activity as appropriate.

Fig. 13-1 Activities involved in establishing an installation planning schedule (Courtesy, IBM Corp.)

these courses should be notified well in advance, and courses should be presented at times when personnel are free of their normal duties or when there is no interference with normal job functions.

Physical environment: In conjunction with the computer center manager, the systems team should analyze the physical layout of the equipment available or to be installed and the work stations of the employees who will be carrying out functions within the system. Equipment suppliers' engineers should be asked to check power supply, air conditioning, hazardous conditions, etc., and make recommendations to the computer center manager.

Telecommunications: If the system is to have remote data gathering or communications capabilities, arrangments should be made at this time for common carrier lines (communications links) and the various services should be contracted for. In addition to Bell Telephone, several companies provide communications lines, including Western Union.

Selection of software: Besides the computer programs that must be prepared by the organization implementing the system, other software will be needed from the equipment manufacturer or outside sources. Computer programs which perform utility functions and sorts must be decided upon. The operating system to be used should have already been determined by the systems team; however, new factors may cause some adjustment in the selection of the operating system and the software needed for sorts and utility functions.

Systems testing: If the new system is to be run on new equipment that has not yet arrived, then testing should be done on computer time which can be rented perhaps from the equipment supplier. At this time, programs should be coded and run. Adjustment to the programs should continue until the system is working as designed. Test data should be used at first, but actual data must be run through the system before it can be said to work as planned. Programs which are to be leased or purchased, such as sorts, should be integrated into the system at this time.

File conversion: If the system is to use data that has already been prepared but is not in the format of the new system or is not coded the same as the new system, then file conversion must be planned for. There are two ways to convert files to new formats or new codes. First, a computer program can be written which will read the old files as input and produce the new file as output. Secondly, files may be keyboarded using the original source documents. File conversion should be planned so that the files are ready when the system is ready to be implemented.

Ordering supplies: All preprinted forms, tapes, disks, etc., that the system will require should be ordered in time for them to be delivered when needed by the system. Proper storage space for these supplies should be prepared, and the analyst should consider the physical security of both source documents and data files.

Run book preparation: The computer operator's instruction sheets and operating procedures should have been prepared by this time. See Chap. 10 for more information on the Run Book.

Systems implementation: Two methods can be used to begin system operation. These are total or parallel conversion.

Total conversion means that the old system is discontinued and the new system is installed. The disadvantage of this method is that the new system must work correctly or the function which it performs will not be carried out, and the organization will suffer because the output is not produced or is not produced efficiently.

Parallel operation means that the old and the new system run at the same time; therefore, any deficiencies in the new system are "backed up" by the old system. Sometimes the parallel operation can be run for several months until management and systems personnel are satisfied that the newly installed system is meeting its objectives. At this time, the old system is discontinued and the new system is depended upon to carry out its function.

System evaluation: Once the system is operating as expected, the systems team has finished its job. However, periodic evaluation is needed as the new system operates. Sometimes, the systems team leaves checklists for either the computer center manager or other management personnel to check the system's production. Sometimes, the systems personnel will be asked to make a check or audit of the system. Whatever the method used, it is probable that the system will be dynamic (changing) rather than static (no change). Some personnel should be designated as responsible for systems maintenance. This includes the changing of the system as certain functions are noted that could be improved. The systems team will also find it desirable to interview system users perhaps six months to a year later to determine if the system is producing the information or service expected of it.

SUMMARY

Once a system has been analyzed and designed, it must be put into operation. Implementation of the system is sometimes turned over to management personnel, such as the computer center manager.

The systems analyst will be called upon to coordinate the implementation, to act as a consultant, and probably to teach a few training courses on how the system works.

The systems analyst may be involved in ordering equipment, selecting personnel to run the system, selecting of software, systems testing, file conversion, parallel operations, and the evaluation of the system.

Although the systems analyst may go on to other projects while the system is being implemented, he will probably be called upon to check the system after it has run for a few months and also to help adjust the system when problems are noted.

Questions

1. When does the system team leave a system and go on to design some other system?
2. Explain how the systems team may have to coordinate with purchasing or contracting personnel.
3. Make a list of some of the courses you believe would be necessary to train personnel when a new system is installed. Write a short description of each course.
4. List some physical site-planning conditions which should be checked.
5. Name some computer programs that might be purchased by an organization instead of written and coded by in-house programmers.
6. How would you begin systems testing if the computer was not delivered for six months?
7. Explain how you would convert data files that are to be used on a different type or make of computer.
8. Explain parallel conversion.
9. Make a checklist which you believe could be used for systems evaluation.

Appendix A
A Description of the Work Environment of a Systems Analyst

MANAGEMENT SYSTEMS DIVISION

This division is composed of four departments: Design, Development, Operations, and Services. They function as internal management consulting organizations with broad responsibilities for management systems effort in all segments of the company. Active projects are carried on within Manufacturing, Advertising, Buying, Sales, Comptrollers, Marketing, Engineering, and Research & Development Divisions of Procter & Gamble. The division has responsibility for the management systems activities of our foreign operations as well as those within the United States.

Procter & Gamble has long been recognized as a pioneer in the application of traditional industrial engineering concepts, such as time study, wage incentives, materials handling and methods engineering. Today Management Systems is also deeply involved with the techniques of data processing, behavioral and management science, and operations research as applied to the design and operation of complex systems. The latest in computing equipment and a complete technical library are available for this activity.

The division is a relatively small corporate staff organization with about 250 members of management. Approximately one-third have advanced college degrees.

DESIGN DEPARTMENT

This department design's systems to process commercial and technical activities and to provide the information requirements of managers.

Activities include the application of management sciences to projects in these areas, the design and installation of major systems to support corporate transactions, short-term problem solving, and improvement of the effectiveness of the managerial decision process.

This department is divided into client-oriented functions, described individually below:

Supply Systems Responsible for the development of techniques and systems that assist the company departments responsible for the functions of purchasing, raw materials and supplies inventory control, inbound freight handling and accounts payable.

Manufacturing Systems Works closely with plant and central manufacturing management personnel in analytic problem-solving, prototype systems testing, local or corporate systems design and improvement of going systems.

Emphasis is on data gathering, modeling, and optimization techniques which help a manager to control and plan manufacturing operations. The group also works with plants or regional data centers serving plants to help determine needs for telecommunications and computing hardware.

Sales Systems Services the line and staff sales departments of the company. Activities include the application of management sciences to projects in these areas and range through the design and installation of major systems to support sales transactions and managerial decision-making.

Product Management Systems Works with marketing, research, and manufacturing groups to bring to bear techniques of operations research and computer science in the management of company products. Typical marketing problems include promotion evaluation, media allocation, and new item shipment forecasting. Division-wide problems include project evaluation, project management, and multiproduct resource allocation.

Administrative Systems Works with departments that make significant use of financial information as part of their managerial process. Our objective is to continually assist these managers in improving financial systems design, data utilization techniques, and organization effectiveness.

Distribution Systems Services all groups concerned with the distribution function. Here, designers meet the technical and informational demands of the systems involved in a complex network which satisfies finished product supply and demand needs.

DEVELOPMENT DEPARTMENT

This department is mainly composed of people with extensive technical backgrounds concerned with design and equipment development as well as implementation of technical projects. Here, a small staff including Ph.D.-level consultants assist data systems designers by selecting or developing fundamental analytical tools to improve the design process.

This department is responsible for forecasting long-range corporate requirements for computer capabilities and for obtaining or developing the hardware and system control software necessary to meet these requirements.

Applied Research is conducted in quantitative areas involving statistics, operations research, econometrics and psychometrics. Consulting services are also provided in management science, involving such techniques as simulation, linear programming, multiple regression, and decision theory.

OPERATIONS DEPARTMENT

This department represents a vital blend of management and technical consulting talent. It is divided into three groups, described individually below:

Corporate Data Center   The hub of the data processing network is connected via telephone lines to the computers in the regional data centers, and to computers and both high-speed and low-speed terminals in other P&G and subsidiary locations. It is simultaneously handling, in a multiprogrammed environment, batch jobs, remote entry jobs, and on-line inquiry and conversational programming jobs.

Regional Data Processing Network   The regional data processing network consists of nine data centers located in various cities from coast to coast. The computers in the regional data centers send and receive data, primarily between themselves and the Corporate Center, for an extensive commercial system. In addition, they do work for nearby P&G and subsidiary facilities.

Data Center Services   The basic responsibilities of Data Center Services fall into four categories:

1. Short-range hardware planning
2. Complete system software support
3. Consulting services to computer users and to those facilities providing computing services to users
4. Operating three technical computing data centers in the Cincinnati area. In addition, responsibilities for lease-back equipment arrangements and the coordination of the use of outside time-sharing services are fulfilled.

SERVICES DEPARTMENT

This department is largely composed of experienced resource people who provide professional consulting, project coordination, and communication services to functional organizations throughout the company.

The main thrust of the department comes in the areas of managing and promoting change, obtaining and measuring results, and linking analytical, physical, and behavioral considerations for maximum effectiveness. Approaches to development of the individual, the organization, and the management process are studied and applied.

Existing industrial engineering tools are employed, but much effort is devoted to the development of new techniques which will facilitate change in all segments of the business with special attention given to profit improvement.

PHILOSOPHY AND ENVIRONMENT OF MANAGEMENT SYSTEMS

This section and those following are intended to provide the recruiter with information and to give the candidate an idea of what working in this area would really be like.

Management Systems is a challenging area of work in which creativity and resourcefulness play a major role. You become part of a relatively small staff organization whose principal objective is to improve the management process and to satisfy the data processing needs of the company. We continuously search for new and better ways to accomplish objectives.

Four departments comprise our Management Systems division: Design, Development, Operations, and Services. All are located in Cincinnati, corporate headquarters of P&G, and operate as internal consulting and service organizations staffed by persons of diverse managerial backgrounds.

The character of our four Management Systems departments is one of broad diversity in terms of clients served as well as techniques employed. Members have extensive freedom in carrying out their tasks and developing and testing their ideas.

Many different academic concentrations provide the foundation for a starting career in Management Systems: engineering, industrial management, sciences, mathematics, management science, operations research, statistics, computer science, and business administration.

The desire to continue to learn, the ability to express yourself clearly, and flexibility in approaching a problem are important personal qualities. We're seeking individuals who will develop into systems managers early in their careers.

A P&G manager leads, makes decisions, and thinks in broad terms about our business. He anticipates needs, plans what should be done to meet those needs, and makes sure the plans are properly carried out. He uses the kind of imagination in his daily work that creates new and better ideas.

Initial Responsibilities: Since Procter & Gamble's Management Systems activities are headquartered in Cincinnati, most first assignments are there. However, some are in one of the regional data centers, which are located in Trenton, Boston, Baltimore, Chicago, St. Louis, Kansas City, Dallas, Los Angeles, and Sacramento.

The needs of the company, along with your interests and academic or work experience, determine your initial assignment. You could begin in any one of the four departments—Design, Development, Operations or Services—but those with bachelor's or master's degrees usually start in Design or Operations. Individuals with higher degrees in mathematics, statistics, and computer science usually begin in Development. The Services Department is composed mainly of experienced individuals from many parts of the company.

The Kind of Work You Would Be Doing: Training is primarily accomplished through the professional experience gained on projects. You learn by doing, although at some stage in the first few months you may receive up to four weeks of technical training tailored to supplement your background.

P&G believes in early responsibility as the best possible way to learn. So, the company takes your education as a starting point and gives you a series of planned job experience and responsibilities to develop your ideas and abilities.

Some typical projects which might be assigned to you during your early months with P&G are:

Design, install, and maintain a system to predict the future price of certain agricultural commodities used by P&G.

Develop a shipment forecasting and production planning system to help manufacturing managers optimize company costs.

Develop a mathematical model to prescribe the optimum loading pattern to be used for mechanically loading trucks and railcars.

Assume supervisory responsibility in the Corporate Data Center.

Develop a module for the Promotion Planning and Control System to be used by advertising management.

Coordinate planning and installation of a process control computer in a paper products plant.

Development through working experience is continuous and brings you in contact with managers who can help you grow and learn rapidly. You always have the benefit of guidance and counsel from your immediate manager. In addition to project work, you will participate in seminars designed to expand your technical knowledge.

Your Career: P&G follows the practice of promotion from within and advancement is based on merit. We know of no other organization where there is greater opportunity to advance on the basis of merit alone. This means that performance is the primary criterion. It means going beyond your assigned work and reaching out for other ways to contribute. It also means actively seeking additional responsibility and the exercise of leadership by you whenever possible.

We review your progress with you to provide a basis for personal planning. Our steady growth is constantly generating new needs for management in many areas. Because of their corporate nature, the four Management Systems Departments afford excellent opportunities to be actively involved with all phases of the business—advertising, sales, manufacturing, market research, purchasing, engineering, finance, distribution, foreign operations and technical staff.

To continue your own personal development, normal business contacts are supplemented by continuing education programs, many of which are conducted by P&G staff people. Subjects covered include courses such as basic and advanced statistics, matrix algebra, psychometrics, inventory control, cost control, and management. For self-enrichment, members attend outside courses, seminars, and professional meetings.

The knowledge and experience of all members of management is a vital resource—our managers combine their talents and abilities to help you succeed.

Index

Index

DATE DUE

OCT 2 1 1982	
NOV 1 7 1982	
JAN 6 1983	
MAR 9 1984	
DEC 2 8 1984	